Lecture Notes in Physics

Volume 986

The series Lecture Notes in Physics (LNP), founded in 1969, reports new developments in physics research and teaching - quickly and informally, but with a high quality and the explicit aim to summarize and communicate current knowledge in an accessible way. Books published in this series are conceived as bridging material between advanced graduate textbooks and the forefront of research and to serve three purposes:

- to be a compact and modern up-to-date source of reference on a well-defined topic;
- to serve as an accessible introduction to the field to postgraduate students and non-specialist researchers from related areas;
- to be a source of advanced teaching material for specialized seminars, courses and schools.

Both monographs and multi-author volumes will be considered for publication. Edited volumes should however consist of a very limited number of contributions only. Proceedings will not be considered for LNP.

Volumes published in LNP are disseminated both in print and in electronic formats, the electronic archive being available at springerlink.com. The series content is indexed, abstracted and referenced by many abstracting and information services, bibliographic networks, subscription agencies, library networks, and consortia.

Proposals should be sent to a member of the Editorial Board, or directly to the responsible editor at Springer:

Dr Lisa Scalone
Springer Nature
Physics
Tiergartenstrasse 17
69121 Heidelberg, Germany
lisa.scalone@springernature.com

More information about this series at http://www.springer.com/series/5304

Michele Arzano • Jerzy Kowalski-Glikman

Deformations
of Spacetime Symmetries

Gravity, Group-Valued Momenta, and Non-Commutative Fields

 Springer

Michele Arzano
Department of Physics
University of Naples Federico II
Naples, Italy

Jerzy Kowalski-Glikman
Institute for Theoretical Physics
University of Wrocław
Wrocław, Poland

National Centre for Nuclear Research
Swierk, Poland

ISSN 0075-8450   ISSN 1616-6361 (electronic)
Lecture Notes in Physics
ISBN 978-3-662-63095-2   ISBN 978-3-662-63097-6 (eBook)
https://doi.org/10.1007/978-3-662-63097-6

This Springer imprint is published by the registered company Springer-Verlag GmbH, DE part of Springer Nature.
The registered company address is: Heidelberger Platz 3, 14197 Berlin, Germany

*To Ania, Łukasz, Kamil, Szymon,
and Tobiasz.
To the memory of Maria Luna Arzano.*

Preface

This monograph is intended as an introduction, at the graduate level, to deformations of the algebras of spacetime symmetries arising in certain topological regimes of (quantum) gravity coupled to point particles. In these regimes, the momentum space of the particles is described by a non-abelian Lie group, and the curvature of the group manifold plays the role of the deformation parameter. These deformed symmetry models are described by non-trivial Hopf algebras and are intimately linked to non-commutative field theories in which the commutators of spacetime coordinates close a Lie algebra. It is speculated that some predictions of deformed symmetry models, if indeed present in nature, can be experimentally verified even with current or foreseeable observational technologies, making these models relevant for the quantum gravity phenomenology research programme.

In Part I of the book, we focus on the link between gravity and deformed symmetries. After introducing the subject matter of the book in Chap. 1, in Chap. 2, we investigate in some detail the emergence of deformations in the context of gravity in $2 + 1$ dimensions. Chapter 3 is devoted to description of gravity in $3 + 1$-dimensions as a constrained topological BF theory, its coupling to point particles, and to speculations how deformations could arise also in this context.

Part II is devoted to a systematic study classical particles whose momentum space is a group manifold. In Chap. 4, we discuss in detail the structure of the phase space of such system. In Chap. 5, we introduce and describe in some detail the κ-Poincaré Hopf algebra, the most studied deformed algebra of spacetime symmetries associated to group-valued momenta in $3 + 1$ dimensions.

In Part III, we illustrate some basic aspects of classical and quantum field theories on κ-Minkowski non-commutative space, the spacetime counterpart of the momentum group manifold associated to the κ-Poincaré algebra.

Part I is mostly self contained (even though some of the concepts encountered there are treated more extensively in Chap. 5 of Part II) while Part III requires some basic notions which are introduced, again, in Chap. 5.

The content of these notes should be accessible to advanced graduate students with a basic background of general relativity, group theory, and differential geometry. An exception is Chap. 5 where we made an effort to give a pedagogical and physically motivated introduction to Hopf algebras and to the basic structures of the κ-Poincaré Hopf algebra. We also feel that the material collected here could serve as an introduction to deformed spacetime symmetries for young and experienced researchers working in quantum gravity and non-commutative field theory and as a good reference for those active in this very field.

To our knowledge this is the first work in which an attempt is made to collect results in this field which have been accumulating for the past twenty years and can be only found scattered in the literature. Most of the topics touched upon in these notes are at the frontier of research on non-commutative field theory and its relations to (quantum) gravity and we thus decided to give a preference to what appear to us the most firmly established results in the field. Of course this choice is highly idiosyncratic but we made an effort to include references to works where alternative approaches are adopted when possible.

Itri, Italy
Warsaw, Poland
January 2021

Michele Arzano
Jerzy Kowalski-Glikman

Acknowledgements

We would like to thank many people who helped us to understand various aspects of deformed theories, classical and quantum gravity. We are particularly grateful to Giovanni Amelino-Camelia, Angel Ballesteros, Andrzej Borowiec, Jose Manuel Carmona, Francesco Cianfrani, Laurent Freidel, Aurelio Grillo, Giulia Gubitosi, Rob Leigh, Jerzy Lewandowski, Stefano Liberati, Jerzy Lukierski, Joao Magueijo, Antonino Marcianò, Flavio Mercati, Catherine Meusburger, Jakub Mielczarek, Djordje Minic, Francisco Nettel, Daniele Pranzetti, Giacomo Rosati, Bernd Schroers, and Lee Smolin. We thank our past and present graduate students Andrea Bevilacqua, Lennart Brocki, Remik Durka, Michał Szczachor, Tomasz Trzes'niewski, Josua Unger, and Adrian Walkus.

We thank Andrea Bevilacqua for reading the earlier versions of the manuscript and pointing out misprints and inaccuracies and Giulia Gubitosi for providing us with her code to generate figures.

Over the years that led to the completion of this monograph, MA has been supported by the European Commission through a Marie Curie Grant and a Fellowship within the programmes FP7-PEOPLE-2011-CIG and FP7-PEOPLE-IEF-2008. He also benefited from funds from the John Templeton Foundation.

For JK-G, this work was supported by funds provided by the National Science Center, project numbers 2019/33/B/ST2/00050 and 2017/27/B/ST2/01902.

This book contributes to the European Union COST Action CA18108 *Quantum gravity phenomenology in the multi-messenger approach.*

Contents

About the Authors

Michele Arzano is a researcher in theoretical physics at the University of Naples 'Federico II'. He obtained his Ph.D. from the University of North Carolina at Chapel Hill and has held positions at the Perimeter Institute for Theoretical Physics in Canada, at the Institute for Theoretical Physics at Utrecht University in the Netherlands, and at 'Sapienza' University of Rome in Italy. His research interests over the years have ranged from non-commutative field theory and deformed symmetries to quantum effects in curved spacetimes and black hole thermodynamics. More recently, he has contributed to research on asymptotic symmetries in general relativity and conformal quantum mechanics.

Jerzy Kowalski-Glikman is a Professor at the Institute of Theoretical Physics of the University of Wroclaw and at National Centre for Nuclear Research in Warsaw. He got his Ph.D. in 1985 from the University of Warsaw and habilitation in 1994 from the University of Wroclaw and full professorship in 2002. His research interests include quantum gravity and quantum gravity phenomenology, non-commutative field theories, cosmology, and string theory. He is the Author of more than hundred research papers published on international peer-reviewed journals.

Part I

From Gravity to Curved Momentum Space and Non-commutative Spacetime

Invitation: Gravity, Point Particles, and Group-Valued Momenta

1

1.1 Introduction

The growth of our understanding of physics of the fundamental constituents of matter in the last century is truly remarkable. Hundred years ago, general relativity had been just formulated, and the researchers still had to wait to witness the dawn of quantum mechanics, to mention only these two major theories that shaped the development of physics ever since.

In spite of the fact that these two great theories, separately, have been extremely successful in describing all known aspects of the physical world, their unification seems still elusive and the formulation of a theory of *quantum gravity* is regarded, for decades now, as the single most important challenge facing modern theoretical physics. This theory, when developed, would provide a missing link between gravity and quantum, which is necessary to complete the 'unfinished revolution' [1] of the XXth century physics. Unfortunately, the quantum gravity research programme faces not only the well-known tremendous technical and conceptual difficulties, but also the seemingly complete lack of experimental feedbacks.

About 20 years ago, the possibility of developing a 'quantum gravity phenomenology' research programme started to take shape [2]. It was noticed that in spite of the fact that the direct observation of quantum gravity effects like the scattering of elementary particles at Planckian energies or the direct detection of a single graviton (see [3]) are well beyond the reach of any foreseeable technology, there may still exist phenomena of quantum gravity origin that might be observable as a result of the possible presence of powerful signal *amplifiers*. For example, the minute effect of an interaction of freely moving particles with the quantum "spacetime foam" may get amplified to an observable size if the time of flight of the particle is long enough (for the recent discussion of the phenomenological consequence of this effect, see [4].)

© Springer-Verlag GmbH Germany, part of Springer Nature 2021
M. Arzano and J. Kowalski-Glikman, *Deformations of Spacetime Symmetries*,
Lecture Notes in Physics 986,
https://doi.org/10.1007/978-3-662-63097-6_1

It has been rather clear from the very first days of the quantum gravity phenomenology research programme that the most promising class of effects to look
for are those associated with possible modifications of spacetime symmetries of special relativity. Indeed, the flat Minkowski space (or spaces that are flat to a good
approximation at the scales of interest) is a configuration of the gravitational field
characterized by the ten-parameter Poincaré group of global spacetime symmetries.
It is feasible that there might be effects of quantum gravitational origin that do not
vanish in the case when the coarse-grained spacetime is Minkowski. Then it is rather
natural to expect that these effects may somehow alter the symmetries of Minkowski
spacetime embodied in the Poincaré group.

These were the motivations that served as a launching pad for two major research
projects: one positing departures from Lorentz invariance, leading to the construction
of models with Lorentz Invariance Violation (LIV) (see [5–7] for reviews), and
the second, known under the name of Doubly (or Deformed) Special Relativity
(DSR), in which a milder alteration of Poincaré symmetries was proposed. In both
cases, it was implicitly understood that deviations from the standard symmetries of
special relativity are of quantum gravity origin, although the explicit relation relations
between the two was never understood in a satisfactory way.

Doubly Special Relativity [8,9] was based on the intuition that since diffeomorphism and local Lorentz invariances play such a central role in both classical and
quantum gravity, the Minkowski space should still possess a ten-parameter group
of spacetime symmetries and the relativity principle should hold even if quantum
gravity effects are taken into account. These effects (or some remnants of them)
exhibit themselves through the presence of an additional observer-independent scale
of dimension of mass (or length), which can be identified with the Planck mass M_{Pl}
(or Planck length ℓ_{Pl},) which are related to *deformations* of the Poincaré symmetry
algebra. Soon after the original papers [8,9] Appeared, some explicit models of DSR
were proposed [10–13]. The interested reader will find more detailed information
about DSR in the reviews [14–16].

Although DSR was devised to be a rather general scheme, the so-called κ-Poincaré
algebra [17–20], proposed some years earlier, served as inspiration and provided a
major example of the DSR model from the very beginning. This is a Hopf algebra
with deformation parameter κ of dimension of mass, which is expected to be of the
order of Planck mass.[1] We will discuss the κ-Poincaré algebra in more detail in the
next chapter, but let us just mention its most important feature, namely that contrary
to the standard Poincaré symmetry of special relativity; in the case of the κ-Poincaré
symmetry, the action of both translations and Lorentz transformation depends on
the momentum (and spin) of the state it acts on. For example, the worldlines of
two particles with different momenta are translated, according to κ-Poincaré, by a
different, momentum-dependant amounts, which means that the two worldlines may
cross for a local observer but miss each other for a translated one.

---

[1] Since the value of the deformation parameter is to be derived from some fundamental theory and/or
experiments, in what follows we will use κ to denote the deformation parameter, whose value is
not fixed, while the term 'Planck mass' will refer to $M_{Pl} = \sqrt{\hbar/G} \sim 10^{19}$ GeV.

This leads to an apparent conflict with the locality principle, noticed first in [21] and further discussed in [22,23]. In the papers [24,25], the lack of absolute locality was elevated to the status of a new principle, called *The principle of Relative Locality*, which states that when quantum gravity effects are taken into account, locality loses its absolute status and becomes relative. It was also realized that the relative locality is a direct consequence of a non-trivial geometry of momentum space.

The idea that momentum space might be curved is quite old. It seems that it was first spelled out by Max Born in the paper [26], where it is argued that some kind of 'reciprocity principle' should be adopted, stating that both curved spacetime and curved momentum space should be involved simultaneously in the description of (quantum) physics. About ten years later in the seminal paper [27], Snyder argued that curvature in momentum space might be necessary to handle ultraviolet divergencies of quantum field theory. This paper introduces, as a bi-product, a non-commutativity of spacetime coordinates and a minimal length, arguing that both do not need to be in conflict with Lorentz symmetry (for a recent review of minimal length scenarios, see Ref. [28].) The ideas of Snyder were later expanded by several Soviet groups, most notably by Kadyshevsky et al. (see Ref. [29] and references therein.). In the context of DSR models, it was observed [36,37] that the κ-Poincaré algebra can be naturally understood in terms of the momentum space being the four-dimensional Lie group $AN(3)$, which as a manifold is a sub-manifold of four-dimensional de Sitter space. This showed that symmetry deformation provides a concrete realization of the old ideas of curved momentum space dating back to the first intuitions of Max Born.[2] We will return to this construction in the next chapter.

In all these attempts, the introduction of curved momentum space had a purely utilitarian character: it was aimed at solving some outstanding problems and/or providing a novel technical perspective. The question arises however is there any fundamental reason to believe that the momentum space is actually curved? Although the complete story is not known, one can give an argument, ultimately relating curved momentum space with the theory of quantum gravity.

The argument is based on the intuition that the presence of a scale is a prerequisite for the emergence of a non-trivial manifold. This intuition was beautifully expressed by Carl Friedrich Gauss already at the dawn of differential geometry:

> The assumption that the sum of the three angles [of a triangle] is smaller than 180° leads to a geometry which is quite different from our (Euclidean) geometry, but which is in itself completely consistent. I have satisfactorily constructed this geometry for myself [...], except for the determination of one constant, which cannot be ascertained a priori. [...] Hence I have sometimes in jest expressed the wish that Euclidean geometry is not true. For then we would have an absolute a priori unit of measurement.[3]

---

[2] In the recent years, Born's ideas found their new incarnations in investigations of new mathematical structures in symplectic geometry (see [30] and references therein) and in a new model in string theory, called the metastring theory [31–35].

[3] As cited in [38].

The necessity of the presence of a scale is easy to understand. Indeed any non-trivial geometry requires nonlinear structures and those can be constructed only if there is a scale that makes it possible to construct nonlinear expressions from fundamental, dimensionful variables. One can interpret Gauss' dictum as the statement that if a scale of some physical quantity is present in a theory, one could expect that the geometry of the corresponding manifold must be non-trivial. Or putting it in other words: 'everything is curved unless it cannot be [because the scale is not available.]'

There are several examples that support this claim. The closest to our considerations comes from relativistic kinematics. In Galilean relativity, there is no room for a non-trivial geometry of the velocity space, because the velocity scale is not available. There is therefore no alternative to the linear composition of velocities

$$\mathbf{v} \oplus \mathbf{u} = \mathbf{v} + \mathbf{u}.$$

Special relativity introduces a scale of velocity, the velocity of light[4] c, and according to Gauss' dictum, one would be led to suspect that the manifold of (three) velocities possesses non-trivial structures. And indeed it does. Contrary to Galilean mechanics, in special relativity, the velocity composition law is highly non-trivial

$$\mathbf{v} \oplus \mathbf{u} = \frac{1}{1 + \mathbf{u}\mathbf{v}/c^2} \left(\mathbf{v} + \frac{\mathbf{u}}{\gamma_v} + \frac{1}{c^2} \frac{\gamma_v}{1 + \gamma_v} (\mathbf{v}\mathbf{u})\mathbf{v} \right) , \quad \gamma_v = \sqrt{1 - \mathbf{v}^2/c^2}. \quad (1.1)$$

This expression is neither symmetric nor associative. Indeed there is quite non-trivial mathematics behind it (see [39] and references therein) which has interesting physical consequences like the Thomas precession.

The relativistic four-momentum space is, arguably, even more important physically than spacetime itself. Indeed virtually all physical measurements can be reduced to measurements of energies and momenta of incoming particles of various kinds (probes) performed by measuring devices located at the origin of a coordinate system, and therefore we deal with momentum space, not spacetime, measurements. It is only by observing the incoming probes that we can infer the properties of distant events [24,25]. Therefore, momentum space measurements should be seen as physically more fundamental than spacetime measurements. (*I don't see space...I see [images of] things* the renowned Mexican painter Diego Rivera used to say.) A natural question is whether we have good reasons to believe that momentum space is described by an almost structureless Minkowski space or if it is conceivable that it could posses more intricate geometrical structures.

Following Gauss' intuition, a possible way of addressing this question is to look for a theory that could provide us with a fundamental momentum scale κ. Such a theory indeed exists. In $2 + 1$ spacetime dimensions, Newton's constant G has dimensions of an inverse mass raising the hope that it may provide the sought momentum scale being a prerequisite for the emergence of a non-trivial momentum space geometry. This is indeed the case and will be discussed in some detail in Sect. 1.2 below.

---

[4]In this book, we use the units in which $c = 1$.

What about gravity in the physical $3 + 1$ dimensions? In this case, Newton's constant is the ratio of Planck length l_{Pl} and the Planck mass M_{Pl}

$$l_{Pl} = \sqrt{\hbar G}, \quad M_{Pl} = \sqrt{\frac{\hbar}{G}}, \tag{1.2}$$

and therefore has the dimension of length over mass. Therefore, the full quantum regime of 3+1-dimensional gravity, where both G and $\hbar$ are relevant can be characterized by the fact that the length scale of the system (process) is of the order of Planck length *and* the relevant energy scale is of the order of Planck mass. However, one can imagine a regime of quantum gravity, in which the Planck length is negligible, while the Planck mass remains finite. This formally means that both $\hbar$ and G go to zero, so that both quantum and local gravitational effects become negligible, while their ratio remains finite [24,25,40]. In more physical terms, this regime is realized if the characteristic length scales relevant for the processes of interest are much larger than l_{Pl}, so that the spacetime quantum foamy effects can be safely neglected, while the characteristic energies are comparable with the Planck mass. An example of such a process might be the gravitational scattering in the case when the longitudinal momenta are Planckian, while the transferred momentum is very small (as compared to M_{Pl}) [41,42]. In the case of such processes, we again encounter the situation that the momentum scale is present, and thus we could expect to find a non-trivial geometry of the momentum space. Unfortunately, to date no specific model of this kind has been formulated.

1.2   2+1 Gravity, Particles, and Curvature of Momentum Space

As mentioned above, Newton's constant in 2+1-dimensional gravity has the dimension of inverse mass and therefore one can expect, according to Gauss' *dictum*, that the momentum space in this theory has a non-trivial geometry. Therefore, gravity in 2+1 spacetime dimensions may serve as an archetypal example of a theory with deformed spacetime symmetries. It is remarkable that, contrary to the 3+1-dimensional situation, the emergence of the geometry of momentum space is in 2+1 dimension a purely classical effect, since in order to introduce the mass scale, we do not need to introduce the Planck constant $\hbar$. This fact renders the 2+1-dimensional systems much easier to investigate than the 3+1-dimensional one. Let us discuss, in qualitative terms, how this deformation emerges, leaving the more detailed discussion to the next chapter.

General relativity in 2+1 dimensions has the remarkable feature of not possessing any local degrees of freedom [43]. Indeed the vacuum Einstein equations

$$R_{\mu\nu} = 0 \tag{1.3}$$

force the Ricci tensor to vanish. In 3+1 dimensions, the vanishing of the Ricci tensor does not imply that the spacetime is (locally) flat: indeed in this case, the Ricci tensor has $4 \times 5/2 = 10$ independent components, while the curvature tensor has $(1/12)16 \times 15 = 20$ components. In 2+1 dimension, however, the number of components of the Ricci tensor is $3 \times 4/2 = 6$ which is exactly the number of the components of the Riemann tensor $(1/12)9 \times 8$. In fact, since in 2+1 dimensions the Weyl tensor

$$C_{\mu\nu\rho\sigma} = R_{\mu\nu\rho\sigma} - \left(g_{\mu\rho} R_{\nu\sigma} + g_{\nu\sigma} R_{\mu\rho} - g_{\nu\rho} R_{\mu\sigma} - g_{\mu\sigma} R_{\nu\rho} \right)$$
$$+ \frac{1}{2} R \left(g_{\mu\rho} g_{\nu\sigma} - g_{\nu\rho} g_{\mu\sigma} \right)$$

being totally antisymmetric is identically zero, implying

$$R_{\mu\nu\rho\sigma} = \left(g_{\mu\rho} R_{\nu\sigma} + g_{\nu\sigma} R_{\mu\rho} - g_{\nu\rho} R_{\mu\sigma} - g_{\mu\sigma} R_{\nu\rho} \right)$$
$$- \frac{1}{2} R \left(g_{\mu\rho} g_{\nu\sigma} - g_{\nu\rho} g_{\mu\sigma} \right) . \tag{1.4}$$

This shows that in 2+1 dimensions, the Riemann tensor is completely determined by the Ricci tensor, and since in vacuum the latter is zero, the spacetime is (locally) flat. Thus, in 2+1 dimensions, there are no gravitational waves (and no gravitons), and Newtonian interactions between point masses do not exist.[5]

Since in 2+1 dimensions the vacuum Einstein equations force the Riemann curvature to vanish, any solution of these equations must take the form of flat Minkowski patches, appropriately glued together along the boundaries. It follows that if the topology of spacetime is trivial, a globally flat Minkowski space is the only possible configuration of the classical gravitational field.

Similarly, if the cosmological constant does not vanish, Einstein equations take the form

$$R_{\mu\nu} - \frac{1}{2} g_{\mu\nu} R + \Lambda g_{\mu\nu} = 0. \tag{1.5}$$

Then

$$R = 6\Lambda, \quad R_{\mu\nu} = 2\Lambda g_{\mu\nu},$$

and

$$R_{\mu\nu\rho\sigma} = \Lambda \left(g_{\mu\rho} g_{\nu\sigma} - g_{\nu\rho} g_{\mu\sigma} \right),$$

which means that spacetime is maximally symmetric with constant (positive or negative—depending on the sign of Λ) curvature, i.e., a three-dimensional de Sitter or anti-de Sitter space. Again there is here no room for gravitons and local gravitational interactions.

Things become much more interesting if we add point sources, i.e., point particles having mass m and/or spin s, and for simplicity, we will discuss here spinless

---

[5]It is worth noticing in passing that also in 3+1 dimensions the Einstein equations make the components of the Ricci tensor completely fixed by the distribution of matter. Therefore, all the dynamical degrees of freedom of gravity are captured by the Weyl tensor $C_{\mu\nu\rho\sigma}$.

particles only.[6] In this case, the energy–momentum tensor on the right hand side of the Einstein equations (1.3) becomes proportional to m times a delta function with support on the particle worldline. Recalling (1.4), one easily concludes that the Riemann tensor vanishes everywhere except on the particle worldline, where it has a delta-like singularity, and therefore the geometry of the constant time spaces must be that of a cone. Indeed the metric corresponding to the general solution of Einstein equations with a point-like massive particle (placed at the origin of coordinates) as a source has the form [44]

$$ds^2 = -dt^2 + dr^2 + (1 - 4Gm)^2 r^2 d\phi^2 \,. \tag{1.6}$$

To see that this metric indeed describes the conical spacetime, let us introduce a new variable

$$\tilde{\phi} = (1 - 4Gm)\,\phi \,. \tag{1.7}$$

In terms of this variable the metric has the form

$$ds^2 = -dt^2 + dr^2 + r^2 d\tilde{\phi}^2 \,, \tag{1.8}$$

which is just the metric of the flat 2+1-dimensional Minkowski space. However, since the range of $\tilde{\phi}$ is $(0, 2\pi - 8\pi Gm)$, with the spacetime points with coordinates $(t, r, 0)$ and $(t, r, 2\pi - 8\pi Gm)$ identified, we see that indeed the constant time slices are two-dimensional cones with the *deficit angle* $\alpha = 8\pi Gm$. The full three-dimensional spacetime can be seen as ordinary Minkowski space with a wedge of angular size $\alpha = 8\pi Gm$ removed and the sides identified. The effect of removing a wedge is to create a "jump" in the coordinate ϕ of magnitude α at the points, where the wedge was removed. If we denote by Σ_1 and Σ_2 the two half-planes delimiting the wedge, the gluing condition leading to the conical geometry with deficit angle α requires that the faces should be identified via a rigid rotation by the angle α

$$\Sigma_1 = R(\alpha)\Sigma_2 \,. \tag{1.9}$$

Since the deficit angle α is proportional to the mass of the particle, we can say that *the rest energy of the particle is parametrized by the rotation* $R(\alpha)$. Let us now turn to a moving defect. A direct way to characterize the motion of such defect is via the boost $B(\eta)$ which maps the defect into a stationary one.[7] For the half-planes delimiting the wedge describing the defect, we have

$$\begin{aligned} \Sigma_1' &= B(\eta)\Sigma_1, \\ \Sigma_2' &= B(\eta)\Sigma_2, \end{aligned}$$

---

[6]We will consider spinning particles in the next chapter. Let us remark here that a non-vanishing spin of the point particle is associated to a *time-offset* in the conical geometry. See, e.g., [47] for details.

[7]For the technical details about the procedure of boosting a conical defect, see [45].

Fig. 1.1 The conical space of a massive particle at rest

where Σ_1' and Σ_1' belong to a stationary defect with deficit angle α so that $\Sigma_1' = R(\alpha)\Sigma_2'$. The gluing of the faces of the wedge describing a moving defect is therefore given by

$$\Sigma_1 = B^{-1}(\eta)R(\alpha)B(\eta)\,\Sigma_2\,. \tag{1.10}$$

In analogy with the case of stationary defect, the *Lorentz transformation*

$$B^{-1}(\eta)R(\alpha)\,B(\eta)$$

encodes *all the information needed to describe the motion of the defect*. This provides an intuitive picture of how a curved momentum space emerges in this context: the Lorentz transformation $B^{-1}(\eta)R(\alpha)B(\eta)$ can be seen as the "on-shell" physical momentum associated to a moving point particle coupled to $2 + 1$ gravity with the "kinematical" momentum space given by the three-dimensional Lorentz group which as a manifold is an anti-de Sitter space, a space with constant negative scalar curvature.

Similar conclusions can be reached using the construction presented in [46] in which the analogies with the description of a point particle in Minkowski space are more transparent. The basic idea is to derive an equation for particle's worldline using the degrees of freedom of gravity.

The first step is to regularize the singular field configuration describing the conical spacetime (see the left panel in Fig. 1.1). In order to do so, one 'cuts off' the tip of the cone (see the middle panel Fig. 1.2) introducing polar coordinates r, ϕ, with $r = 0$ corresponding to the upper circle and the angular coordinate ϕ having the standard range $\phi \in [0, 2\pi]$. To make sure that the geometry of the space depicted in the middle panel is the same as the original one, we impose the condition, called the *particle condition*, that the circumference of the upper circle vanishes, i.e., $g_{\phi\phi}(r=0, \phi)=0$.

At this point, it is convenient to switch from the metric formalism to the frame field one. In this formalism, the fundamental fields describing gravity are the frame field e_μ^a and the connection ω_μ^a. The components of the frame field and the metric are related by

$$g_{\mu\nu} = \eta_{ab}\, e_\mu^a\, e_\nu^b\,,$$

and therefore the particle condition takes the form

$$e_\phi^a(r = 0, \phi) = 0\,. \tag{1.11}$$

Fig. 1.2 The conical space without the tip and the cut at $\phi = 0$

Since there are no sources for the gravitational field on the manifold in the right panel Fig. 1.2, the dynamics of the gravitational field on this manifold is described by vacuum Einstein equations forcing the spacetime curvature and torsion to vanish. It follows that on any simply connected region of the manifold

$$\omega_\mu^a = (\mathfrak{l}^{-1} \partial_\mu \mathfrak{l})^a \,, \quad e_\mu^a = (\mathfrak{l}^{-1} \partial_\mu \mathbf{q} \,\mathfrak{l})^a \,, \tag{1.12}$$

where $\mathfrak{l}$ is the (spacetime-dependent) element of the three-dimensional Lorentz group SO(2, 1) generated by J^a, while $\mathbf{q} = q^a \, P_a$ a field valued in $R^{2,1}$, which we identify, as a vector space, with the abelian Lie algebra generated by three commuting translation generators P_a of the 2+1-dimensional Poincaré group (see the next chapter for details). The map $\mathbf{q}$ can be interpreted geometrically as an *embedding* of the simply connected portion of spacetime into Minkowski space while the group-valued field $\mathfrak{l}$ maps the local frames in the conical spacetime to a reference frame in the embedding Minkowski space [46].

The second step of the construction consists in turning the regularized cone into a simply connected manifold so that the solution to the field equations (1.12) is defined globally on it. To this end, we cut the manifold along the line $\phi = 0$. The resulting space is now simply connected and the solution is globally defined on it. There is however the price to pay, namely we must impose a continuity conditions along the cut, so that

$$\omega_\mu^a(t, r, \phi = 0) = \omega_\mu^a(t, r, \phi = 2\pi) \,, \quad e_\mu^a(t, r, \phi = 0) = e_\mu^a(t, r, \phi = 2\pi) \,. \tag{1.13}$$

Fortunately, these continuity conditions can be easily solved by assuming that the fields on both sides of the cut are gauge equivalent

$$\mathfrak{l}(t, r, \phi = 0) = u^{-1} \mathfrak{l}(t, r, \phi = 2\pi) \,, \quad \mathbf{q}(t, r, \phi = 0) = u^{-1} \left(\mathbf{q}(t, r, \phi = 2\pi) - \mathbf{v} \right) u \,, \tag{1.14}$$

where $u \in$ SO(2, 1) and $\mathbf{v} \in \mathfrak{so}(2, 1)$ are constant elements of the Lorentz group and the algebra, respectively.

Let us concentrate now on the 'particle boundary' at $r = 0$. It follows from (1.11) and (1.12) that the function $\mathbf{q}$ is only a function of time $\mathbf{q} = \mathbf{x}(t)$. In fact we, introduce here such a suggestive notation to stress that the **q** field at the particle boundary represents the particle's *worldline* in the embedding space. In order to find the differential equation obeyed by $\mathbf{x}(t)$, we notice that according to (1.14) we must have

$$\mathbf{x}(t) - \mathfrak{u}\,\mathbf{x}(t)\,\mathfrak{u}^{-1} = \mathbf{v}\,. \tag{1.15}$$

Taking time derivative of this equation, we see that $\dot{\mathbf{x}}(t)$ must be invariant under the adjoint action of $\mathfrak{u}$. This is only possible if $\mathfrak{u}$ is a group element of the form of the exponent of a Lie algebra vector proportional to the velocity $\dot{\mathbf{x}}(t)$. To see the physical implication of this, let us parametrize the group element $\mathfrak{u}$ in terms of 2x2 matrices as

$$\mathfrak{u} = P_3 \mathbb{1} + \frac{1}{\kappa}\, P_a\, J^a\,, \tag{1.16}$$

where $\mathbb{1}$ is the identity matrix, J^a a basis of $\mathfrak{so}(2, 1)$

$$[J_a, J_b] = \varepsilon_{abc}\, J^c\,, \quad a, b, c = 0, 1, 2\,,$$

and we introduced the Planck mass $\kappa \equiv (4\pi G)^{-1}$ so that the components P_a have the canonical mass dimension. Since $\dot{\mathbf{x}}(t)$ is invariant under the adjoint action of $\mathfrak{u}$, we must have that

$$\mathbf{x}(t) = \mathbf{y} + \frac{1}{\kappa}\tau(t)\mathbf{P}\,, \tag{1.17}$$

where $\mathbf{y}$ is a constant vector in $\mathfrak{so}(2, 1)$ and $\tau(t)$ a real function of time. Equation (1.17) can be seen as the parametric equation for the worldline of a particle passing through the point $\mathbf{y}$ and having *momentum* $\mathbf{p}$.

Now we come to our main point: since $\mathfrak{u}$ is an element of the group SO(2, 1), its determinant must be equal to 1, and this implies that

$$P_3^2 - \frac{1}{4\kappa^2}\, \eta^{ab}\, P_a\, P_b = 1\,, \quad \eta^{ab} = (-1, 1, 1)\,. \tag{1.18}$$

Such equation is the quadric defining a three-dimensional Anti-de Sitter space, a space with constant negative curvature. Therefore, the functions P_a can be seen as *coordinates* on a curved *momentum space*, which in this case is simply the manifold of the Lorentz group SO(2, 1).

Equation 1.18 describes a deformation of the particle kinematics, resulting in the fact that the momentum space, being a 2+1-dimensional linear space (or an abelian group manifold) becomes a group manifold of the group SO(2, 1), with a non-trivial geometry. The inverse Newton's constant κ, of the dimension of mass, becomes a deformation parameter. When the Newton's constant goes to zero, and the effects of gravity are being completely switched off, $\kappa \to \infty$ and the deformation disappears.

The argument presented strongly suggest that the momentum space of such particle/defects should be curved. In the next chapter, we will give a rigorous proof of this fact using the Chern–Simons formulation of three-dimensional gravity.

References

1. Rovelli, C.: Unfinished revolution. In: Oriti, D. (Ed.) Approaches to Quantum Gravity. Cambridge University Press (2009). arXiv:gr-qc/0604045
2. Amelino-Camelia, G.: Are we at the dawn of quantum gravity phenomenology? Lect. Notes Phys. **541**, 1 (2000). arXiv:gr-qc/9910089
3. Dyson, F.: Is a graviton detectable? Int. J. Mod. Phys. A **28**, 1330041 (2013)
4. Vasileiou, V., Granot, J., Piran, T., Amelino-Camelia, G.: A Planck-scale limit on spacetime fuzziness and stochastic Lorentz invariance violation. Nat. Phys.
5. Mattingly, D.: Modern tests of Lorentz invariance. Living Rev. Rel. **8**, 5 (2005). arXiv:gr-qc/0502097
6. Liberati, S., Mattingly, D.: Lorentz breaking effective field theory models for matter and gravity: theory and observational constraints. arXiv:1208.1071 [gr-qc]
7. Bluhm, R.: Observational Constraints on Local Lorentz Invariance. arXiv:1302.1150 [hep-ph]
8. Amelino-Camelia, G.: Testable scenario for relativity with minimum length. Phys. Lett. **B 510**, 255 (2001). arXiv:hep-th/0012238
9. Amelino-Camelia, G.: Relativity in space-times with short distance structure governed by an observer independent (Planckian) length scale. Int. J. Mod. Phys. **D 11**, 35 (2002). arXiv:gr-qc/0012051
10. Kowalski-Glikman, J.: Observer independent quantum of mass. Phys. Lett. **A 286**, 391 (2001). arXiv:hep-th/0102098
11. Bruno, N.R., Amelino-Camelia, G., Kowalski-Glikman, J.: Deformed boost transformations that saturate at the Planck scale. Phys. Lett. **B 522**, 133 (2001). arXiv:hep-th/0107039
12. Magueijo, J., Smolin, L.: Lorentz invariance with an invariant energy scale. Phys. Rev. Lett. **88**, 190403 (2002). arXiv:hep-th/0112090
13. Magueijo, J., Smolin, L.: Generalized Lorentz invariance with an invariant energy scale. Phys. Rev. **D 67**, 044017 (2003). arXiv:gr-qc/0207085
14. Kowalski-Glikman, J.: Introduction to doubly special relativity. Lect. Notes Phys. **669**, 131 (2005). arXiv:hep-th/0405273
15. Kowalski-Glikman, J.: Doubly special relativity: facts and prospects. In: Oriti, D. (Ed.) Approaches to Quantum Gravity. Cambridge University Press (2009). arXiv:gr-qc/0603022
16. Amelino-Camelia, G.: Doubly-special relativity: facts, myths and some key open issues. Symmetry **2**, 230 (2010). arXiv:1003.3942 [gr-qc]
17. Lukierski, J., Ruegg, H., Nowicki, A., Tolstoi, V.N.: Q deformation of Poincaré algebra. Phys. Lett. **B 264**, 331 (1991)
18. Lukierski, J., Nowicki, A., Ruegg, H.: New quantum Poincaré algebra and k deformed field theory. Phys. Lett. **B 293**, 344 (1992)
19. Lukierski, J., Ruegg, H., Zakrzewski, W.J.: Classical quantum mechanics of free kappa relativistic systems. Ann. Phys. **243**, 90 (1995). arXiv:hep-th/9312153
20. Majid, S., Ruegg, H.: Bicrossproduct structure of kappa Poincaré group and noncommutative geometry. Phys. Lett. **B 334**, 348 (1994). arXiv:hep-th/9405107
21. Hossenfelder, S.: Bounds on an energy-dependent and observer-independent speed of light from violations of locality. Phys. Rev. Lett. **104**, 140402 (2010). arXiv:1004.0418 [hep-ph]
22. Amelino-Camelia, G., Matassa, M., Mercati, F., Rosati, G.: Taming nonlocality in theories with Planck-Scale deformed Lorentz symmetry. Phys. Rev. Lett. **106**, 071301 (2011). arXiv:1006.2126 [gr-qc]
23. Smolin, L.: Classical paradoxes of locality and their possible quantum resolutions in deformed special relativity. Gen. Rel. Grav. **43**, 3671 (2011). arXiv:1004.0664 [gr-qc]
24. Amelino-Camelia, G., Freidel, L., Kowalski-Glikman, J., Smolin, L.: The principle of relative locality. Phys. Rev. **D 84**, 084010 (2011). arXiv:1101.0931 [hep-th]
25. Amelino-Camelia, G., Freidel, L., Kowalski-Glikman, J., Smolin, L.: Relative locality: a deepening of the relativity principle. Gen. Rel. Grav. **43**, 2547 (2011) [Int. J. Mod. Phys. D **20**, 2867 (2011)]. arXiv:1106.0313 [hep-th]

26. Born, M.: A suggestion for unifying quantum theory and relativity. Proc. R. Soc. Lond. **A 165**, 291 (1938)
27. Snyder, H.S.: Quantized space-time. Phys. Rev. **71**, 38 (1947)
28. Hossenfelder, S.: Minimal Length Scale Scenarios for Quantum Gravity. arXiv:1203.6191 [gr-qc]
29. Kadyshevsky, V.G., Mateev, M.D., Mir-Kasimov, R.M., Volobuev, I.P.: Equations of motion for the scalar and the spinor fields in four-dimensional noneuclidean momentum space. Theor. Math. Phys. **40**, 800 (1979) [Teor. Mat. Fiz. **40**, 363 (1979)]
30. Freidel, L., Rudolph, F.J., Svoboda, D.: A unique connection for Born geometry. Commun. Math. Phys. **372**(1), 119–150 (2019). https://doi.org/10.1007/s00220-019-03379-7, arXiv:1806.05992 [hep-th]
31. Freidel, L., Leigh, R.G., Minic, D.: Born reciprocity in string theory and the nature of spacetime. Phys. Lett. B **730**, 302 (2014). arXiv:1307.7080
32. Freidel, L., Leigh, R.G., Minic, D.: Quantum gravity, dynamical phase space and string theory. Int. J. Mod. Phys. D **23**(12), 1442006 (2014). arXiv:1405.3949 [hep-th]
33. Freidel, L., Leigh, R.G., Minic, D.: Metastring theory and modular space-time. JHEP **06**, 006 (2015). https://doi.org/10.1007/JHEP06(2015)006, arXiv:1502.08005 [hep-th]
34. Freidel, L., Leigh, R.G., Minic, D.: Phys. Rev. D **94**(10), 104052 (2016). https://doi.org/10.1103/PhysRevD.94.104052, arXiv:1606.01829 [hep-th]
35. Freidel, L., Kowalski-Glikman, J., Leigh, R.G., Minic, D.: Theory of metaparticles. Phys. Rev. D **99**(6), 066011 (2019). https://doi.org/10.1103/PhysRevD.99.066011, arXiv:1812.10821 [hep-th]
36. Kowalski-Glikman, J.: De sitter space as an arena for doubly special relativity. Phys. Lett. **B 547**, 291 (2002). arXiv:hep-th/0207279
37. Kowalski-Glikman, J., Nowak, S.: Doubly special relativity and de Sitter space. Class. Quant. Grav. **20**, 4799 (2003). arXiv:hep-th/0304101
38. Milnor, J.: Hyperbolic geometry: the first 150 years. Bull. Am. Math. Soc. **6**, 9 (1982)
39. Girelli, F., Livine, E.R.: Special relativity as a non commutative geometry: lessons for deformed special relativity. Phys. Rev. **D 81**, 085041 (2010). arXiv:gr-qc/0407098
40. Girelli, F., Livine, E.R., Oriti, D.: Deformed special relativity as an effective flat limit of quantum gravity. Nucl. Phys. **B 708**, 411 (2005). arXiv:gr-qc/0406100
41. 't Hooft, G.: Graviton dominance in ultrahigh-energy scattering. Phys. Lett. **B 198**, 61 (1987)
42. Verlinde, H.L., Verlinde, E.P.: Scattering at Planckian energies. Nucl. Phys. **B 371**, 246 (1992). arXiv:hep-th/9110017
43. Carlip, S.: Quantum Gravity in 2+1 Dimensions, 276p. University Press, Cambridge, UK (1998)
44. Staruszkiewicz, A.: Acta Phys. Polon. **24**, 735 (1963)
45. van de Meent, M.: Piecewise Flat Gravity in 3+1 Dimensions. arXiv:1111.6468 [gr-qc]
46. Matschull, H.-J., Welling, M.: Quantum mechanics of a point particle in (2+1)-dimensional gravity. Class. Quant. Grav. **15**, 2981 (1998). arXiv:gr-qc/9708054
47. de Sousa Gerbert, P.: On spin and (quantum) gravity in (2+1)-dimensions. Nucl. Phys. B **346**, 440 (1990)

Gravity in 2+1 Dimensions as a Chern–Simons Theory

2

In the previous chapter, we argued that the effective theory of particles coupled to gravity in 2+1 dimensions can be understood as a deformations of the standard relativistic particles theory. Now it is time to present the rigorous derivation of this result. We start with the formulation of the pure 2+1 gravity [1,2] in terms of the Chern–Simons theory (an extensive discussion of gravity in 2+1 dimensions can be found in the book of Carlip [3]). Then we discuss the coupling of particles, following the ideas of [4,5], to be followed by a detailed derivation of the deformed particle action, with curved momentum space, describing the relativistic particle moving in its own gravitational field. We conclude this chapter with a discussion on how deformed spacetime symmetries arise directly from quantum gravity in 2+1 dimensions.

2.1 Pure Gravity in 2+1 Dimensions

It is well known that when the cosmological constant vanishes, gravity can be described as a gauge theory of Poincaré group [6]. Similarly, when the positive (negative) cosmological constant is present, gravity can be formulated as a gauge theory with de Sitter (Anti-de Sitter) gauge group, for positive (negative) cosmological constant, respectively. These three groups are the symmetry groups of three maximally symmetric spaces: Minkowski space ($\Lambda = 0$), Anti-de Sitter space ($\Lambda < 0$), and de Sitter space ($\Lambda > 0$). Thus, gravity emerges from gauging the spacetime symmetries of maximally symmetric spacetimes; one can easily check that if the spacetime dimension is D+1, the number of such symmetries is $N = (D + 1)(D + 2)/2$, which in the case of 2+1 spacetime dimensions, we are interested in, gives $N = 6$. Indeed, if the space has 2 dimensions, we have two independent boosts and one rotation,

© Springer-Verlag GmbH Germany, part of Springer Nature 2021
M. Arzano and J. Kowalski-Glikman, *Deformations of Spacetime Symmetries*,
Lecture Notes in Physics 986,
https://doi.org/10.1007/978-3-662-63097-6_2

which together forms the three-dimensional Lorentz group, with generators J_a.[1] Similarly, we have three independent spacetime translations and we denote the three independent translation generators by P_a.

The generators of the gauge group collectively denoted as $T_I = (J_a, P_a)$ satisfy the following commutational algebra

$$[T_I, T_J] = f_{IJ}{}^K T_K . \tag{2.1}$$

In the case of the gauge algebras of 2+1 gravity, we have

$$[J_a, J_b] = \epsilon_{ab}{}^c J_c , \quad [J_a, P_b] = \epsilon_{ab}{}^c P_c , \quad [P_a, P_b] = -\Lambda \epsilon_{ab}{}^c J_c , \tag{2.2}$$

where $\Lambda > 0$ corresponds to de Sitter, $\Lambda < 0$ to anti-de Sitter, and $\Lambda = 0$ to Poincaré algebra. Notice the crucial fact that in 2+1 dimensions, the number of Lorentz generators is equal to the number of translational ones. This coincidence happens only in 2+1 dimensions and makes 2+1 gravity special: some constructions that we are going to employ here are not available in any other dimension.

In the gauge theory of gravity, the gravitational field is not described by the metric tensor $g_{\mu\nu}$ (which, as we will see in a moment, in this formulation is not fundamental but composite), but instead by the gauge field $A_\mu(x)$, valued in an appropriate gauge algebra (2.2). Therefore, we have

$$A_\mu(x) \equiv A_\mu^I(x) T_I = \omega_\mu^a(x) J_a + e_\mu^a(x) P_a . \tag{2.3}$$

Usually the components of $A_\mu(x)$ in this decomposition $\omega_\mu^a(x)$ and $e_\mu^a(x)$ are called, respectively, Lorentz connection and triad or dreibein (tetrad or vierbein in 3+1 dimensions). They have the following geometric interpretation. The triad $e_\mu^a(x)$ is the mapping of vectors $V^\mu(x)$ belonging to the tangent space of the curved manifold at point x, to vectors $V^a(x)$ belonging to the single ambient flat Minkowski space. One often says that the triad makes it possible to trade the 'curved' index μ for the 'flat' one a and vice versa. Such mapping is not unique, it is defined up to a point-dependent Lorentz transformation, and $\omega_\mu^a(x)$ is the gauge field associated with this local symmetry. The relation between the triad and local translations and its geometrical meaning is a bit more subtle and we will discuss it below.

The components of the triad are related to the metric tensor in the usual way

$$g_{\mu\nu}(x) = \eta_{ab} e_\mu^a(x) e_\nu^b(x) , \quad \eta_{ab} = \text{diag}(-1, 1, 1) . \tag{2.4}$$

Again, one sees that the triad makes it possible to translate from the geometry of flat Minkowski space with the metric η_{ab} to the one of the curved manifold with metric $g_{\mu\nu}(x)$. We use the latter to rise and lower curved indices $\mu, \nu, \ldots$ and the

---

[1] In higher dimensions, for Lorentz group, one conventionally uses the generators J_{ab}, antisymmetric in a, b. In 2+1 dimensions, there is a duality between vectors and antisymmetric tensors, and it is convenient to make use of it defining $J_a = \frac{1}{2} \epsilon_a{}^{bc} J_{bc}$.

former to rise and lower the flat ones a, b, ... One should be quite careful, however, because in the gauge formulation of gravity it is *not* assumed that the triad is invertible, and therefore, contrary to the Einsteinian theory of gravity, it may happen that the metric is not invertible everywhere. In spite of this, it turns out that the gauge formulation of gravity does not differ much from the metric one (see, e.g., [7] for detailed discussion.)

Having the gauge field A_μ, we can calculate the associated gauge field strength (called also the curvature) $F_{\mu\nu}$, which again can be decomposed into Lorentz and translational parts

$$F_{\mu\nu} \equiv F^I_{\mu\nu} T_I = \left(\partial_\mu A^I_\nu - \partial_\nu A^I_\mu + A^J_\mu A^K_\nu f_{JK}{}^I \right) T_I = F^a_{\mu\nu} J_a + T^a_{\mu\nu} P_a \,, \quad (2.5)$$

where $F^a_{\mu\nu}$ is called Lorentz curvature and $T^a_{\mu\nu}$ is called torsion. Explicitly we find

$$F^a_{\mu\nu} = \partial_\mu \omega^a_\nu - \partial_\nu \omega^a_\mu + \epsilon^a{}_{bc} \omega^b_\mu \omega^c_\nu - \Lambda \epsilon^a{}_{bc} e^b_\mu e^c_\nu \equiv R^a_{\mu\nu} - \Lambda \epsilon^a{}_{bc} e^b_\mu e^c_\nu \,, \quad (2.6)$$

where the Lorentz subalgebra valued field strength $R^a_{\mu\nu}$ is called Riemannian curvature, because it is related to the Riemann curvature tensor as follows:

$$R^a_{\mu\nu} = \epsilon^a{}_{bc} e^b_\rho e^c_\sigma R^{\rho\sigma}{}_{\mu\nu} \,. \quad (2.7)$$

The translational component of $F_{\mu\nu}$ is called torsion and has the form

$$T^a_{\mu\nu} = \partial_\mu e^a_\nu - \partial_\nu e^a_\mu + \epsilon^a{}_{bc} \left(\omega^b_\mu e^c_\nu - \omega^b_\nu e^c_\mu \right) \,. \quad (2.8)$$

Since A is a gauge field, by construction, it transforms under local gauge transformations defined by the gauge group element $g(x)$ as follows:

$$A_\mu(x) \rightarrow A^{(g)}_\mu(x) = g^{-1}(x) A_\mu(x) g(x) + g^{-1}(x) \partial_\mu g(x) \,. \quad (2.9)$$

Infinitesimally we can write $g = 1 + \varepsilon^I T_I$ and it follows from (2.9) that

$$\delta A_\mu = \partial_\mu \varepsilon^I T_I + f_{JK}{}^I A^J_\mu \varepsilon^K T_I \,. \quad (2.10)$$

The infinitesimal local gauge transformation can be decomposed, as before, into the infinitesimal Lorentz transformation generated by $\lambda^a(x)$ and the infinitesimal translation generated by $\xi^a(x)$, as follows:

$$\varepsilon^I(x) T_I = \lambda^a(x) J_a + \xi^a(x) P_a \,. \quad (2.11)$$

Under gauge transformations the curvature transforms homogeneously, to wit

$$F_{\mu\nu} \rightarrow F'_{\mu\nu} = g^{-1} F_{\mu\nu} g \,, \quad \delta F_{\mu\nu} = f_{JK}{}^I F^J_{\mu\nu} \varepsilon^K T_I \,. \quad (2.12)$$

The fundamental symmetry of any theory of gravity is the general coordinate invariance intimately related to Einstein's principle of relativity (see [8] for detailed discussion,) which is manifest in the Lagrangian formulation of the theory. Thus, the general coordinate invariance is yet another local symmetry and one may wonder if the symmetries are not independent, but are related in some way. Moreover, there seems to be an apparent clash between the local gauge symmetries we defined above and the symmetries of the Einstein's general relativity, which consists of the general coordinate invariance only. The local Lorentz invariance does not cause any problems, because it disappears when one starts using the metric as the fundamental field and the flat indices are not present in the theory anymore. The local translational symmetry survives however the transition to the metric formulation and seems to be problematic, because as a result, we have twice as many local gauge symmetries, which, if there, would render too many degrees of freedom gauge, and therefore, unphysical. It turns out, however, that there is a deep relation between general coordinate invariance and local translations. This can be seen as follows [1]:

Let us first decompose (2.10) in order to obtain the transformations laws for Lorentz connection and triad

$$\delta_{\lambda,\xi}\omega^a_\mu = \partial_\mu\lambda^a + \epsilon^a{}_{bc}\,\omega^b_\mu\,\lambda^c - \Lambda\,\epsilon^a{}_{bc}\,e^b_\mu\,\xi^c\,, \tag{2.13}$$

$$\delta_{\lambda,\xi}e^a_\mu = \partial_\mu\xi^a + \epsilon^a{}_{bc}\,\omega^b_\mu\,\xi^c + \epsilon^a{}_{bc}\,e^b_\mu\,\lambda^c\,, \tag{2.14}$$

On the other hand, under an infinitesimal general coordinate transformations generated by a vector field ζ^μ these components transform as one-form coefficients, so that

$$\delta_\zeta\omega^a_\mu = \zeta^\nu\partial_\nu\,\omega^a_\mu + \omega^a_\nu\partial_\mu\zeta^\nu\,, \quad \delta_\zeta e^a_\mu = \zeta^\nu\partial_\nu\,e^a_\mu + e^a_\nu\partial_\mu\zeta^\nu\,. \tag{2.15}$$

At the first sight (2.13), (2.14) and (2.15) describe completely different transformations. To see that they are, in fact, closely related, let us consider (2.14) with $\lambda = 0$ and $\xi^a = e^a_\mu\,\zeta^\mu$. Then this equation becomes

$$\delta e^a_\mu = \zeta^\nu\partial_\nu\,e^a_\mu + e^a_\nu\partial_\mu\zeta^\nu + \zeta^\nu\,T^a_{\mu\nu} + \epsilon^a{}_{bc}\,e^b_\mu\,\left(\zeta^\nu\,\omega^c_\nu\right)\,.$$

The first two terms are identical with (2.15), the third term is proportional to torsion, while the last one can be interpreted as a local Lorentz transformation with the parameter $\lambda^c \equiv \zeta^\nu\,\omega^c_\nu$. As we will see in a moment, one of the field equation of 2+1 gravity enforce the torsion tensor to vanish, and we conclude that, on-shell, the local translations of the triad are combinations of a general coordinate transformation and local Lorentz symmetry. The reader may check it by direct calculation that the same conclusion holds for the Lorentzian component of the gauge field, ω^a_μ, the only difference being that instead of the torsion in the resulting equation, we have to do with the curvature $F^a_{\mu\nu}$ (2.6).

Having discussed the basic building blocks of 2+1 gravity, the gauge fields of 2+1-dimensional spacetime algebra (2.2) and their symmetries we can now turn to the construction of the dynamics of the theory. To guess its form, let us first notice that in 3+1 dimensions, in tetrad formalism, the Einstein-Hilbert Lagrangian $L_{3+1} \sim$

$\sqrt{-g}(R + \Lambda)$ can be rewritten as $L_{3+1} \sim \epsilon(e\,e\,R + \Lambda e\,e\,e\,e)$. In 2+1 dimension, the epsilon symbol has three indices, and therefore, we expect something of the sort $L_{2+1} \sim \epsilon(e\,R + \Lambda e\,e\,e)$. The first term looks like $\epsilon\,(e\,\partial\omega + e\,\omega\,\omega)$ and it very much reminds the terms in the Chern–Simons action. Therefore, we postulate that the action of gravity in 2+1 dimensions will have the Chern–Simons form

$$I = \frac{k}{4\pi} \int d^3x\, \epsilon^{\mu\nu\rho} \left\langle A_\mu \partial_\nu A_\rho + \frac{2}{3} A_\mu A_\nu A_\rho \right\rangle. \tag{2.16}$$

which can be also rewritten as

$$I = \frac{k}{4\pi} \int d^3x\, \epsilon^{\mu\nu\rho} \left\langle A_\mu \partial_\nu A_\rho \right\rangle + \frac{1}{3} \left\langle A_\mu \left[A_\nu, A_\rho \right] \right\rangle. \tag{2.17}$$

In this formula, we denote by $\langle \star \rangle$ an inner product on the Lie algebra that makes it possible to contract algebra indices and such that the resulting action is gauge invariant (the action is, of course, manifestly invariant under general coordinate transformation, because $\epsilon^{\mu\nu\rho}$ is a tensorial density and all the spacetime indices are contracted). To see which properties the inner product must satisfy, let us consider an arbitrary gauge transformation (2.10). One checks that the Lagrangian density is gauge invariant (up to a total derivative) if the inner product is symmetric $\langle T_I\, T_J \rangle = \langle T_J\, T_I \rangle$, cyclic $\langle T_I\, T_J\, T_K \rangle = \langle T_K\, T_I\, T_J \rangle$, and ad-invariant $\langle [T_I, T_J]\, T_K \rangle + \langle T_J\, [T_I, T_K] \rangle = 0$.

As shown in [1], in 2+1 dimensions, two such inner products exist. The first is the diagonal one that works in any spacetime dimensions. In this product, we pair Lorentz and translational generators with themselves, to wit

$$\langle J_a\, J_b \rangle = \eta_{ab}, \quad \langle P_a\, P_b \rangle = \Lambda\eta_{ab}, \quad \langle J_a\, P_b \rangle = 0. \tag{2.18}$$

The problem with this product is that it becomes degenerate when the cosmological constant vanishes. This does not necessarily mean that the theory constructed with the help of the inner product (2.18) could not have the smooth vanishing cosmological constant limit.[2] Unfortunately, in this case, it does: in spite of the fact that the field equations do have the smooth $\Lambda \to 0$ limit (in fact they are identical to those that we will derive for another inner product below), the basic Poisson brackets do not, and thus the Hamiltonian formulation of this theory is not well defined in this limit. For this reason, we will not use the inner product (2.18) in what follows.

Another inner product that can be constructed in 2+1 dimensions makes use of the fact that in this dimension the number of Lorentz and translations generators is exactly the same, so that they can be paired with one another, to wit

$$\langle J_a\, P_b \rangle = \sqrt{\Lambda}\,\eta_{ab}, \quad \langle P_a\, P_b \rangle = 0, \quad \langle J_a\, J_b \rangle = 0. \tag{2.19}$$

---

[2]In the next chapter, we will see that in 3+1 dimensions, the theory gravity which uses the analogue of (2.18) is perfectly well defined for all values of the cosmological constant.

Using this inner product in (2.16), we obtain the action

$$I = \frac{\bar{k}}{4\pi} \int d^3x \, \epsilon^{\mu\nu\rho} \, e_{a\mu} \, R^a_{\mu\nu} - \frac{\Lambda}{3} \, \epsilon^{\mu\nu\rho} \, \epsilon_{abc} \, e^a_\mu \, e^b_\nu \, e^c_\rho , \tag{2.20}$$

where

$$R^a_{\mu\nu} \equiv \partial_\nu \omega^a_\rho - \partial_\rho \omega^a_\nu + \epsilon^a{}_{bc} \, \omega^b_\nu \, \omega^c_\rho .$$

In fact, the action (2.20) is nothing but the standard Einstein–Hilbert action of 2+1-dimensional gravity, written in terms of the triad and connection variables, if one identifies

$$\frac{\bar{k}}{4\pi} = \frac{1}{8\pi G} = \frac{\kappa}{2} . \tag{2.21}$$

where $\kappa \equiv (4\pi G)^{-1}$ is the three-dimensional Planck mass. Thus, $\bar{k}$ has the dimension of mass, contrary to the coupling constant k in (2.17), which is dimensionless. This difference results from the fact that the triad e^a_μ is canonically dimensionless, while the connection A_μ has the dimension of inverse length. Noticing the presence of $\sqrt{\Lambda}$ in (2.19), we see that

$$\frac{k}{4\pi} = \frac{\bar{k}}{4\pi\sqrt{\Lambda}} = \frac{\kappa}{2\sqrt{\Lambda}} . \tag{2.22}$$

The field equations following from this action force both the curvature $R^a_{\mu\nu}$ and torsion $T^a_{\mu\nu}$ to vanish, and thus as shown in Chap. 1, the theory described by them is a topological field theory without local degrees of freedom.

To complete this short description of free 2+1-dimensional gravity, let us return to the action (2.16) and rewrite it in the hamiltonian form that is going to be useful later. To this end, let us assume that the spacetime manifold $\mathcal{M}$ can be decomposed into the real time axis and the spacial manifold, $\mathcal{M} = \mathbb{R} \times \mathcal{S}$. Then the gauge field A_μ has the components A_0, A_i, and in terms of them, the action (2.16) takes the form (after integration by parts)

$$I = \frac{k}{4\pi} \int d^3x \, \epsilon^{ij} \langle \partial_0 A_i A_j \rangle + \epsilon^{ij} \langle A_0 F_{ij} \rangle , \tag{2.23}$$

where the spacial components of the field strength F_{ij} are defined as (cf. (2.5))

$$F_{ij} = \left(\partial_i A^I_j - \partial_j A^I_i + A^J_i \, A^K_j \, f_{JK}{}^I \right) T_I . \tag{2.24}$$

In the action (2.23), A_0 plays the usual role of Lagrange multiplier enforcing the constraint $F_{ij} = 0$. This constraint again says that the (space part of) curvature vanishes, and it follows that on any simply connected region of the space manifold Σ, the potential A_i is pure gauge. We will make use of this, below, in deriving the form of the effective deformed particle action.

One last ingredient which we will need in what follows is the Poisson bracket of the theory which can be immediately derived from (2.23) and reads

$$\left\{A_i^I(x), A_j^J(y)\right\} = -\frac{4\pi}{k} \epsilon_{ij} G^{IJ} \delta^2(x - y). \tag{2.25}$$

where G^{IJ} is the metric associated with the inner product $\langle \star \rangle$.

This completes the description of pure gravity in 2+1 dimensions. Let us now turn to the coupling of gravity to point particles.

2.2 Particle Coupling

Let us start considering a particle moving in spacetime. At each point in spacetime, its motion is characterized by the particle's position and momentum[3] (we assume for a moment that the particle is spinless.) The momentum of the moving particle can be obtained from that of the particle at rest, with components $p_{rest} = (m, 0, 0)$, where m is the particle's rest mass, by applying an appropriate boost, symbolically $p_{moving} = \Lambda p_{rest} = (p_0, p_1, p_2)$. Of course, in the case of a particle in some external field, whose momentum is not constant, the appropriate boost may change from point to point, but still, it exists at any point of the trajectory. Similarly, by applying an appropriate space translation, we can move the particle initially resting at an arbitrary point, to the origin of the coordinate system. Thus, using the boost, we can turn the moving particle to be at rest and then translate it to the origin, and reversing the argument, we conclude that the dynamics of the particle can be fully described by its rest mass m, the (local) Lorentz transformation, and the (local) translation. But the latter are the gauge symmetries of gravity we just have constructed! Thus, we expect that there is a close relation between gauge (naively, unphysical) degrees of freedom of gravity at the position of the particle and dynamical (physical) degrees of freedom of the particle. Sometimes such degrees of freedom of gravity are called 'would be gauge'. Let us see how this general intuition can be precisely formulated. In what follows, we will present the formalism developed in [4,5].

In the previous section, we showed that the gravitational field is described by the Lie algebra valued gauge field A_μ. Being a vector field, such potential can be easily integrated along the particle's worldline. Indeed, if the worldline of the particle is described by the functions $z^\mu(\tau)$, then we can contract the spacetime index of the gauge field with the worldline tangent vector and form the expression $A_z(\tau) \equiv A_\mu(z(\tau)) \dot{z}^\mu$ on it. Further, to get a scalar quantity (without Lie algebra indices) before integrating $A_z(\tau)$ along the worldline, we must contract it with some gauge algebra element, using the inner product $\langle * \rangle$ described above. We also have to specify somehow the particle's properties.

---

[3]Here and below momentum means relativistic momentum, whose components are energy and linear momentum.

These two problems can be solved simultaneously. According to the discussion at the beginning of this section, we can start considering the particle at rest. Then the (normalized) tangent vector to the particle trajectory has the components $\dot{z}^\mu = (1, 0, 0)$ and $A_z = A_0$. At rest, the particle is completely described by its invariant mass m and spin s, characterizing its momentum and internal angular momentum at the rest state. Let us represent them as an element of the Cartan subalgebra (i.e., roughly speaking, the maximal commuting subalgebra of the gauge algebra), for example

$$\mathscr{C} = m\, J_0 + s\, P_0\,. \tag{2.26}$$

Then the action of the particle at rest has the form

$$I^{(0)}_{part} = \int d\tau\, \langle A_0 \mathscr{C} \rangle\,. \tag{2.27}$$

Notice that this action manifestly breaks all the symmetries (gauge and coordinate invariance) on the particle worldline. But this is exactly what we want, because we expect that the gauge degrees of freedom at the position of the particle are going to become the true dynamical degrees of freedom of the particle. Thus, in order to obtain the action of the particle with an arbitrary dynamics, we should replace A_0 with its arbitrary gauge transformation $A_0^{(h)}$ defined in (2.9). In this way, we obtain

$$I_{part} = \int d\tau\, \langle A_0^{(h)} \mathscr{C} \rangle = \int d\tau\, \langle h^{-1}\dot{h}\,\mathscr{C} \rangle + \langle A_0\, h\, \mathscr{C} h^{-1} \rangle \tag{2.28}$$

$$= \int d\tau\, \langle h^{-1}\dot{h}\,\mathscr{C} \rangle + \int d^3x\, \langle A_0(x)\, h(x)\, \mathscr{C} h(x)^{-1} \rangle \delta(x - z(\tau))\,, \tag{2.29}$$

where in the second line, we write the action in the form that we will find convenient later. The first terms of these expressions are the kinetic particle terms, while the second describe its interaction with the gravitational field.

To see this explicitly, let us take, for simplicity, the zero cosmological constant limit,[4] so that the algebra (2.2) becomes the Poincaré algebra

$$[J_a, J_b] = \epsilon_{ab}{}^c J_c\,, \quad [J_a, P_b] = \epsilon_{ab}{}^c P_c\,, \quad [P_a, P_b] = 0\,. \tag{2.30}$$

There are two explicit representations of the Poincaré algebra (2.30) that we are going to use in the calculations below in this chapter. The first is the four-dimensional and can be constructed as follows:

An arbitrary Poincaré gauge group element h can be written as a pair consisting of the Lorentz transformation, represented by the 3×3 orthogonal matrix $\mathbf{l} \in SO(2, 1)$

---

[4]We will stay within this limit for the major remaining part of this chapter.

and the vector $\mathbf{q} \in \mathbb{R}^3$, $h = (\mathbf{l}, \mathbf{q})$ representing translation.[5] The group multiplication and the inverse take the form

$$h_1 h_2 = (\mathbf{l}_1, \mathbf{q}_1)(\mathbf{l}_2, \mathbf{q}_2) = (\mathbf{l}_1 \mathbf{l}_2, \mathbf{q}_1 + \mathbf{l}_1 \mathbf{q}_2), \quad h^{-1} = (\mathbf{l}^{-1}, -\mathbf{l}\mathbf{q}). \tag{2.31}$$

An element of the Poincaré group and algebra can be then represented in the form of 4×4 matrices. We have

$$h = (\mathbf{l}, \mathbf{q}) = \begin{pmatrix} l^a{}_b & q^a \\ 0 & 1 \end{pmatrix}, \quad h^{-1} = (\mathbf{l}^{-1}, -\mathbf{l}^{-1}\mathbf{q}) = \begin{pmatrix} (l^{-1})^a{}_b & -(l^{-1})^a{}_b q^b \\ 0 & 1 \end{pmatrix},$$
$$\tag{2.32}$$

where $l^a{}_b$ is a 3×3 matrix of $SO(2, 1)$. In the same representation the generators of the Poincaré algebra[6] (2.30) are

$$J_a = \begin{pmatrix} -(\epsilon_a)^b{}_c & 0 \\ 0 & 0 \end{pmatrix}, \quad P_a = \begin{pmatrix} 0 & (\delta_a)^b \\ 0 & 0 \end{pmatrix} \tag{2.33}$$

Using this representation, one can easily calculate and explicit expression for Maurer–Cartan form $h^{-1}dh$, where d is any derivative (partial derivative, variation, etc.)

$$h^{-1}dh = \begin{pmatrix} (l^{-1})^a{}_c & -(l^{-1})^a{}_c q^c \\ 0 & 1 \end{pmatrix} \begin{pmatrix} dl^c{}_b & dq^c \\ 0 & 0 \end{pmatrix}$$
$$= \begin{pmatrix} (l^{-1})^a{}_c \, dl^c{}_b & (l^{-1})^a{}_c dq^c \\ 0 & 0 \end{pmatrix} = \omega^a \, J_a + e^a \, P_a, \tag{2.34}$$

where $\omega^a = -\frac{1}{2} \epsilon^a{}_b{}^c \, (l^{-1})^b{}_d \, dl^d{}_c$ and $e^a = (l^{-1})^a{}_c \, dq^c$. We deliberately use here ω^a and e^a to denote the Lorentz and translational component of $h^{-1}dh$, because they are just the connection and the triad one-forms in the case when the curvature and torsion vanish and we have to do with the pure gauge configuration.

Returning to the representation $h = (\mathbf{l}, \mathbf{q})$ we can write, equivalently

$$h^{-1}dh = (\mathbf{l}^{-1}d\mathbf{l}, \mathbf{l}^{-1}d\mathbf{q}). \tag{2.35}$$

There is another, two-dimensional, representation of the Poincaré group that proves particularly convenient in the derivation of the deformed particle Lagrangian that we are going to present below. This representation has been constructed in [13], and has the virtue of unifying the description of (Anti) de Sitter and Poincaré groups and algebras.

---

[5] We refer the reader to Chap. 5 for a pedagogic introduction to semi-direct product groups and the Poincaré group.

[6] We will sometime use $(J_a)^b{}_c = -(\epsilon_a)^b{}_c$ to denote the 3×3 matrix representation of the generators of the Lie algebra $\mathfrak{so}(2, 1)$. It will be clear from the context, which representation we have in mind.

In this representation, the Lorentz generators J_a are identified with the generators of the SL(2, $\mathbb{R}$) group

$$J_0 = \frac{1}{2} \begin{pmatrix} 0 & 1 \\ -1 & 0 \end{pmatrix}, \quad J_1 = \frac{1}{2} \begin{pmatrix} 0 & 1 \\ 1 & 0 \end{pmatrix}, \quad J_2 = \frac{1}{2} \begin{pmatrix} 1 & 0 \\ 0 & -1 \end{pmatrix}, \tag{2.36}$$

satisfying

$$J_a J_b = \frac{1}{4} \eta_{ab} \mathbb{1} - \frac{1}{2} \epsilon_{ab}{}^c J_c . \tag{2.37}$$

The idea now is to represent the translational generators by the same matrices, multiplied by a formal parameter θ, whose interpretation we will find in a moment

$$P_a = \theta J_a . \tag{2.38}$$

Now let us assume that θ^2 is a pure number (although θ *is not*), then

$$[J_a, J_b] = \epsilon_{ab}{}^c J_c , \quad [J_a, P_b] = \epsilon_{ab}{}^c P_c , \quad [P_a, P_b] = \theta^2 \epsilon_{ab}{}^c J_c . \tag{2.39}$$

Comparing this with (2.2), we see that when one identifies θ^2 with $-\Lambda$, one obtains the (Anti) de Sitter algebra. Moreover, by taking $\theta^2 = 0$, we get the Poincaré algebra. Such a construction may seem odd at the first sight, but it makes perfect mathematical sense, see [13] for detailed discussion.

It turns out that any element of the Poincaré group can be represented as a product of the elements of Lorentz and translational groups, generated by J_a and P_a, respectively

$$h = \mathsf{q} \mathfrak{l}, \quad \mathfrak{l} \in \mathrm{SL}(2, \mathbb{R}), \quad \mathsf{q} \in \theta \mathrm{SL}(2, \mathbb{R}) \tag{2.40}$$

Then $h^{-1} = \mathfrak{l}^{-1} \mathsf{q}^{-1}$ and

$$h^{-1} dh = \mathfrak{l}^{-1} \left(\mathsf{q}^{-1} d\mathsf{q} \right) \mathfrak{l} + \mathfrak{l}^{-1} d\mathfrak{l} = \mathfrak{l}^{-1} d\mathsf{q} \mathfrak{l} + \mathfrak{l}^{-1} d\mathfrak{l}, \tag{2.41}$$

where the last equality follows from the fact that $\theta^2 = 0$, so that $\mathsf{q} = \mathbb{1} + \theta q^a J_a$. Equations (2.41) and (2.35) are, of course, consistent with each other

$$\mathfrak{l}^{-1} d\mathsf{q} \mathfrak{l} = \mathfrak{l}^{-1} d\mathsf{q} , \tag{2.42}$$

and we can use the two or four-dimensional representations interchangeably, employing the one that is more convenient in the particular context.

Having all the necessary technical tools, we now return to the action (2.28) and write it down more explicitly, in the case of a free particle, without gravitational interactions. Before doing that, let us stop for a moment, to understand what this last statement means. If we compare the formula (2.29) with (2.23), we see that the second term in the particle action modifies the constraint imposed on the gravitational field strength. Removing this term means that we are turning off the interaction of the particle with the gravitational field. Therefore, in discussing the case of a free particle, we omit the second term in (2.28).

Using formula (2.35) and the properties of the inner product (2.19), we find

$$I^{free} = \int d\tau \, \langle h^{-1}\dot{h}\,\mathscr{C} \rangle = -\int d\tau \, \left(m \, (\mathbf{l}^{-1})^0{}_c \dot{q}^c - \frac{1}{2}\, s \, \epsilon^0{}_a{}^b \, (\mathbf{l}^{-1})^a{}_c \, \dot{\mathbf{l}}^c{}_b \right).$$
(2.43)

Consider the spinless case $s = 0$ first. Since $\mathbf{l}$ is an element of $SO(2, 1)$, $(\mathbf{l}^{-1})^0{}_c$ must satisfy the condition $\eta^{ab} \, (\mathbf{l}^{-1})^0{}_a \, (\mathbf{l}^{-1})^0{}_b \equiv \eta^{00} = -1$ so that we can identify these elements of the Lorentz transformation matrix with the relativistic three-velocity $(\mathbf{l}^{-1})^0{}_a = v_a$. Then mv_a is nothing but the relativistic momentum of the particle p_a, which, by construction, must satisfy the mass-shell condition $\eta^{ab} \, p_a \, p_b = -m^2$. It is convenient (and customary) to impose this condition adding a constraint term to the Lagrangian with the Lagrange multiplier N. Thus, the first term in our action, which describes the linear motion of the particle, reads

$$I_m^{free} = \int d\tau \, p_a \dot{q}^a + N \left(\eta^{ab} \, p_a \, p_b + m^2 \right).$$
(2.44)

Before turning to the spin term, let us make an important observation concerning the 'linear motion' term in the action. As it stands in (2.43), the form of this term is valid in any spacetime dimension. In 2+1 dimensions, there is, however, an alternative equivalent form of this term, which is particularly illuminating, because the backreaction of the particle's gravitational field can be most straightforwardly understood as a deformation of it. To see this, consider the identity

$$\left(\mathbf{l}J_d\mathbf{l}^{-1}\right)^a{}_b = (\mathbf{l}^{-1})_d{}^c \, (J_c)^a{}_b, \quad \text{or} \quad \mathbf{l}^{-1}J_d\mathbf{l} = (\mathbf{l}^{-1})_d{}^c \, J_c,$$
(2.45)

i.e. $m \, \mathbf{l}J_0\mathbf{l}^{-1} = m \, (\mathbf{l}^{-1})^0{}_c \, J_c = p_c \, J^c$. Therefore, in complete agreement with our discussion above, the momentum is indeed the Lorentz transformation acting on the *Lie algebra* element $m \, J_0$, where the Lorentz group acts by conjugation on its Lie algebra. As we will see that the deformation resulting from the presence of the gravitational field will replace this rule with the conjugate action on the *group* element $e^{m/\kappa \, J_0}$.

Let us now turn to the spin term. To understand its meaning and to show that it indeed describes spin of the particle, let us return to the action (2.43) and calculate the equations of motion following from it. Varying over q^a, we find that momentum is conserved $\dot{p}_a = 0$.

Let us then compute the variation over $\mathbf{l}$. To this end, we will need some simple identities. Let us define $\varpi^a{}_b$ to be a 3×3 matrix such that $\delta\mathbf{l}\,\mathbf{l}^{-1} = \varpi$. It follows that the matrix ϖ_{ab} is antisymmetric. Then

$$\delta\mathbf{l} = \varpi \, \mathbf{l}, \quad \delta\mathbf{l}^{-1} \equiv -\mathbf{l}^{-1} \, \delta\mathbf{l}\,\mathbf{l}^{-1} = -\mathbf{l}^{-1} \, \varpi.$$

The second term in (2.43) can, therefore, be rewritten as

$$-\frac{1}{2}\, s \int d\tau \, \langle J_0 \mathbf{l}^{-1} \, \dot{\mathbf{l}} \rangle.$$

Then, using the identities above, its variation can be computed to be

$$-\frac{1}{2}s\int d\tau \left\langle \mathbf{l}\, J_0\, \mathbf{l}^{-1}\, \dot{\varpi} \right\rangle .$$

On the other hand, the variation of $\mathbf{l}$ in the first term in (2.43) gives

$$-\int d\tau\, m\, \delta(\mathbf{l}^{-1})^0{}_c\, \dot{q}^c = -\int d\tau\, m\, (\mathbf{l}^{-1})^{0c}\, \dot{q}^d\, \varpi_{cd} .$$

Collecting these variations and using the fact that by equations of motion the momentum satisfies $\dot{p}_a = 0$, we find the condition that the total angular momentum of the particles M^{ab} is conserved

$$\frac{d}{d\tau} M^{ab} = \frac{d}{d\tau}\left(p^a\, q^b - p^b\, q^a + S^{ab}\right) = 0 , \tag{2.46}$$

where the spin tensor S^{ab} turns out to be

$$S^{ab} = -s\left(\mathbf{l}\, J_0\, \mathbf{l}^{-1}\right)^{ab} . \tag{2.47}$$

It can be easily checked that the spin tensor satisfies the following identities:

$$p_a\, S^{ab} = 0 , \quad S_{ab}\, S^{ab} = 2\, s^2 . \tag{2.48}$$

so that indeed the spin tensor S^{ab} describes the intrinsic angular momentum of moving particle.

This completes our discussion of the undeformed relativistic particle. Let us now turn to the deformations induced by the gravitational field.

2.3   Effective Deformed Single Particle Action

In the previous sections, we reviewed gravity in 2+1 spacetime dimensions and a group-theoretical formulation of the theory of a free point particle with mass and spin and its gravity coupling. Here, following [9, 10], we show how the effective action of a particle coupled to its own gravitational field can be derived. The construction presented in those papers led to an effective symplectic structure of deformed particles. Below we present a more physically appealing but completely analogous construction, which results in a deformed action of the particle.

Before turning to the calculations, let us explain what we can expect to obtain as a result. The Lagrangian of the particle interacting with the gravitational field is a sum of the gravitational (2.23) and the particle one (2.29)

$$L = \frac{k}{4\pi}\int d^2x\, \epsilon^{ij}\, \langle \dot{\mathsf{A}}_i\, \mathsf{A}_j\rangle - \langle h^{-1}\dot{h}\, \mathscr{C}\rangle$$
$$+ \int d^2x\, \left\langle \mathsf{A}_0\left(\frac{k}{2\pi}\epsilon^{ij}\, \mathsf{F}_{ij} - h(x)\, \mathscr{C} h(x)^{-1}\delta^2(\mathbf{x})\right)\right\rangle , \tag{2.49}$$

where the overdot denotes time derivative.

The second term in (2.49) is expressed in terms of a group element h belonging to the relevant (Poincaré, de Sitter or Anti-de Sitter) gauge group defined on the worldline of the particle that encodes all the dynamical information about the particle. On the other hand, away from the particle, the constraint $F_{12} = 0$ following from the condition $\epsilon^{ij} F_{ij} = 0$ forces the connection A_i encoding all the information about the gravitational field to be gauge trivial. Therefore, also, the gravitational field is expressed purely in terms of a group element γ defined on the space manifold. It turns out that h and γ can be combined into a single group element which specifies the deformed particle dynamics. This element describes the particle along with the backreaction of the (topological) gravitational degrees of freedom on the particle's ones, resulting in the effective action that turns out to be a deformed counterpart of the ones derived in the preceding section (2.43).[7] To put it differently, the would be gauge degrees of freedom of gravity on the worldline combine with the dynamical degrees of freedom of the free particle, resulting in the deformed particle dynamics. On the other hand, the degrees of freedom of gravity outside the worldline are still pure gauge and remain physically irrelevant. Let us see how this qualitative ideas can be realized in practice.

The construction that we are to present below becomes a technical nightmare in the case of the (Anti) de Sitter gauge group, and has been done only perturbatively, in the leading orders of small Λ. For this reason, here we will consider only the case of vanishing cosmological constant and Poincaré gauge group, relying on some of the explicit constructions presented in the preceding section.

To proceed let us solve the constraint equation enforced by the Lagrange multiplier A_0

$$\frac{k}{2\pi}\epsilon^{ij} F_{ij} = h \mathscr{C} h^{-1} \delta^2(\mathbf{x}) \tag{2.50}$$

and substitute the solution back to the Lagrangian (2.49) (in the path integral language we would say that we consider an effective action obtained by integrating the field A_0 out), i.e., into

$$I = \frac{k}{4\pi}\int d^{2x}dt\,\epsilon^{ij}\langle \dot{A}_i\, A_j\rangle - \int d\tau\,\langle h^{-1}\dot{h}\,\mathscr{C}\rangle. \tag{2.51}$$

In order to solve the Eq. (2.50), let us decompose the space manifold $\mathscr{S}$ into two subregions (see Fig. 2.1): the plaquette $\mathscr{D}$ being a circle with the centre at the position of the particle, on which we introduce coordinates $0 \geq r \geq 1, 0 \geq \phi \geq 2\pi$, and the asymptotic region $\mathscr{H}$ with $r \geq 1$. These two regions have a common boundary Γ

---

[7]Similar construction has been presented in the papers [11,12], in which one uses the path integral (spin foam) formalism to integrate out the gravitational degrees of freedom in the 2+1-dimensional gravity—scalar field system. As a result, one obtains the deformed action for the scalar field on a non-commutative spacetime.

$A^{\mathcal{H}}$

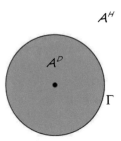

$A^{\mathcal{D}}$

Γ

Fig. 2.1 The decomposition of the manifold

with $r = 1, 0 \geq \phi \geq 2\pi$. On the asymptotic region, the connection is flat and takes
the form

$$A_i^{\mathcal{H}} = \gamma^{-1}\partial_i\gamma\,, \tag{2.52}$$

where γ is an element of the gauge group. One can find the general solution of (2.50)
on the disc as well, it reads

$$A_i^{\mathcal{D}} = \frac{1}{k}\bar{\gamma}^{-1}\mathscr{C}\bar{\gamma}\partial_i\phi + \bar{\gamma}^{-1}\partial_i\bar{\gamma}\,, \quad \bar{\gamma}(0) = h^{-1}\,. \tag{2.53}$$

Using the identity $\epsilon^{ij}\partial_i\partial_j\phi = 2\pi\,\delta^2(\mathbf{x})$, which can be proved by integrating both
sides on an arbitrary disc centred at $\mathbf{x} = 0$, one can easily check that the gauge field
(2.53) indeed satisfies the constraint (2.50).

Let us now plug these solutions back to the action (2.51). Consider first the
asymptotic region $\mathcal{H}$ and decompose the group element γ into the product of a
Lorentz $\mathfrak{l} \in SL(2, \mathbb{R})$ and q in the translational part as in (2.40), $\gamma = \mathfrak{q}\mathfrak{l}$. Then we
find (cf. (2.41))

$$\gamma^{-1}\partial_i\gamma = \mathfrak{l}^{-1}\partial_i\mathfrak{q}\mathfrak{l} + \mathfrak{l}^{-1}\partial_i\mathfrak{l}\,. \tag{2.54}$$

Then we substitute (2.54) into the first term in (2.51). Since the scalar product
is off-diagonal, only the pairings of the terms in Lorentz subalgebra $\mathfrak{l}^{-1}\partial_i\mathfrak{l}$ and the
translational ones $\mathfrak{l}^{-1}\partial_i\mathfrak{q}\,\mathfrak{l}$ contribute. We find that the resulting terms can be written
as a sum of the total time derivative and the total space derivative

$$\epsilon^{ij}\langle\dot{A}_i^{\mathcal{H}}\,A_j^{\mathcal{H}}\rangle = \epsilon^{ij}\partial_0\langle\partial_i\mathfrak{q}\,\partial_j\mathfrak{l}\,\mathfrak{l}^{-1}\rangle + 2\epsilon^{ij}\partial_i\langle\mathfrak{l}\,\mathfrak{l}^{-1}\partial_j\mathfrak{q}\rangle\,. \tag{2.55}$$

Disregarding the total time derivative term, which is not relevant for our argument,
we integrate the total space derivative term, obtaining a boundary term

$$L^{\mathcal{H}} = \frac{k}{2\pi}\int_{\mathcal{H}}d^2x\,\partial_i\left\langle\epsilon^{ij}\,\mathfrak{l}^{-1}\mathfrak{i}\,\mathfrak{l}^{-1}\partial_j\mathfrak{q}\,\mathfrak{l}\right\rangle = \frac{k}{2\pi}\int_{\Gamma}d\phi\left\langle\mathfrak{l}^{-1}\mathfrak{i}\,\mathfrak{l}^{-1}\frac{\partial\mathfrak{q}}{\partial\phi}\,\mathfrak{l}\right\rangle\,. \tag{2.56}$$

Now we can turn to the more complicated contribution to the Lagrangian (2.51)
coming from the disc $\mathcal{D}$. In this case, the expression for the connection (2.53) has
two terms. The term in the Lagrangian resulting from two $\bar{\gamma}^{-1}\partial\bar{\gamma}$ is identical in

form to the expression (2.56); the contribution from the product of two second terms vanishes because $\epsilon^{ij} \partial_i \phi \, \partial_j \phi = 0$. We are, therefore, left with

$$\frac{1}{2\pi} \int_{\mathscr{D}} d^2x \, \epsilon^{ij} \left(\partial_0 (\bar{\gamma}^{-1} \mathscr{C} \bar{\gamma} \partial_i \phi) \bar{\gamma}^{-1} \partial_j \bar{\gamma} \right) , \qquad (2.57)$$

where we again neglected the total time derivative. This expression can be rewritten as a combination of the total time derivative and exact differential, to wit

$$\left\langle \epsilon^{ij} \partial_0 (\bar{\gamma}^{-1} \mathscr{C} \bar{\gamma} \partial_i \phi) \bar{\gamma}^{-1} \partial_i \bar{\gamma} \right\rangle =$$

$$\partial_0 \left\langle \epsilon^{ij} \mathscr{C} \partial_i \phi \, \bar{\gamma} \partial_j \bar{\gamma}^{-1} \right\rangle + \partial_i \left\langle \epsilon^{ij} \mathscr{C} \partial_j \phi \, \dot{\bar{\gamma}} \, \bar{\gamma}^{-1} \right\rangle - 2\pi \, \delta(\mathbf{x}) \left\langle \mathscr{C} \, \bar{\gamma} \dot{\bar{\gamma}}^{-1} \right\rangle . \qquad (2.58)$$

Since $\bar{\gamma}^{-1}(0) = h$ the last term in this expression neatly cancels the second term in the Lagrangian (2.51). Neglecting again the total time derivative in (2.58), collecting all the contributions and adopting the opposite orientation of Γ for the terms coming from the disc $\mathscr{D}$, we see that the Lagrangian (2.51) contains the boundary terms only and has the form

$$L = \frac{k}{2\pi} \int_{\Gamma} d\phi \left\langle \Gamma^{-1} \mathfrak{l} \, \Gamma^{-1} \frac{\partial \mathfrak{q}}{\partial \phi} \mathfrak{l} - \bar{\Gamma}^{-1} \dot{\bar{\mathfrak{l}}} \bar{\Gamma}^{-1} \frac{\partial \bar{\mathfrak{q}}}{\partial \phi} \bar{\mathfrak{l}} \right\rangle - \frac{1}{2\pi} \int_{\Gamma} d\phi \left\langle \mathscr{C} \dot{\bar{\gamma}} \, \bar{\gamma}^{-1} \right\rangle . \qquad (2.59)$$

Now there is time to make a decisive step. We must take into account the fact that the connection A_i is to be continuous across the boundary Γ, i.e.

$$A_i^{\mathscr{H}} \bigg|_{\Gamma} \equiv \gamma^{-1} \partial_i \gamma \big|_{\Gamma} = \frac{1}{k} \bar{\gamma}^{-1} \mathscr{C} \bar{\gamma} \partial_i \phi + \bar{\gamma}^{-1} \partial_i \bar{\gamma} \bigg|_{\Gamma} \equiv A_i^{\mathscr{D}} \bigg|_{\Gamma} , \qquad (2.60)$$

which can be solved to give

$$\gamma|_{\Gamma} = \eta(t) \exp \left(\frac{1}{k} \mathscr{C} \phi \right) \bar{\gamma} \bigg|_{\Gamma} , \qquad (2.61)$$

where $\eta(t) = \mathfrak{e}(t) \mathfrak{n}(t)$ with $\mathfrak{n} \in SL(2, \mathbb{R})$, $\mathfrak{e} \in SL(2, \mathbb{R})$, is an arbitrary element of the Poincaré group depending only on time.

It is worth pausing here for a moment to discuss the meaning of this result. An important thing to observe is that the Lie algebra element $\mathscr{C}$ describing the particle's mass and spin, i.e., its momentum and angular momentum at rest was replaced by the corresponding Lie group element, obtained by exponentiation of the former. On the other hand, the Lagrangian (2.59), to which we will substitute (2.61) completely describes the particle and its gravitational field. These observations suggest that in the final Lagrangian the momentum and spin of the particle are going to be described by group elements, not the algebra ones (although in the case of spin this does not make any real difference because the translational part of the Poincaré group is abelian.) As we will see in a moment this is exactly what is going to happen.

Let us return to the continuity condition. From the translational part of (2.60), we infer that

$$\mathfrak{l}^{-1}\frac{\partial \mathfrak{q}}{\partial \phi}\,\mathfrak{l} = \bar{\mathfrak{l}}^{-1}\frac{\partial \bar{\mathfrak{q}}}{\partial \phi}\bar{\mathfrak{l}} + \frac{1}{k}\bar{\mathfrak{l}}^{-1}\bar{\mathfrak{q}}^{-1}\,\mathscr{C}\bar{\mathfrak{q}}\,\bar{\mathfrak{l}}, \tag{2.62}$$

which simplifies the Lagrangian (2.59) to the form

$$L = \frac{k}{2\pi}\int_{\Gamma} d\phi\left\{u^{-1}\dot{u}\left(\frac{\partial \bar{\mathfrak{q}}}{\partial \phi} - \frac{1}{k}\bar{\mathfrak{q}}^{-1}\mathscr{C}\bar{\mathfrak{q}}\right)\right\}, \tag{2.63}$$

where we introduced a new variable $u \equiv \mathfrak{l}\bar{\mathfrak{l}}^{-1}$ and dropped a term that was a total time derivative.

In order to simplify the expression (2.63) further, we make use of the Lorentz part of the continuity condition (2.61)

$$\mathfrak{l} = \mathfrak{n} \exp\left(\frac{1}{k}\mathscr{C}_J\phi\right)\bar{\mathfrak{l}} \tag{2.64}$$

which gives us an equation for u

$$u = \mathfrak{l}\bar{\mathfrak{l}}^{-1} = \mathfrak{n} \exp\left(\frac{1}{k}\mathscr{C}_J\phi\right). \tag{2.65}$$

Substituting this expression to the Lagrangian (2.59), we find

$$L = \frac{k}{2\pi}\int_{\Gamma}d\phi\,\frac{\partial}{\partial \phi}\left\langle e^{-\frac{1}{k}\mathscr{C}_J\phi}\dot{\mathfrak{n}}^{-1}\mathfrak{n}e^{\frac{1}{k}\mathscr{C}_J\phi}\bar{\mathfrak{q}} - \mathfrak{n}^{-1}\dot{\mathfrak{n}}\,\frac{1}{k}\mathscr{C}_P\phi\right\rangle. \tag{2.66}$$

Integrating (2.66) over ϕ from 0 to 2π and noticing that $\bar{\mathfrak{q}}$ is a single valued function on Γ so that $\bar{\mathfrak{q}}(0) = \bar{\mathfrak{q}}(2\pi)$, we obtain

$$L = \frac{k}{2\pi}\left\langle \dot{\Pi}\,\Pi^{-1}\,\mathfrak{x}\right\rangle - \left\langle \mathscr{C}_P\mathfrak{n}^{-1}\dot{\mathfrak{n}}\right\rangle, \tag{2.67}$$

with $\mathfrak{x} \equiv \mathfrak{n}\bar{\mathfrak{q}}(0)\mathfrak{n}^{-1}$ and the 'group valued momentum' Π is defined as

$$\Pi \equiv \mathfrak{n}e^{-\frac{2\pi}{k}\mathscr{C}_J}\mathfrak{n}^{-1} = \mathfrak{n}e^{-\frac{m}{k}J_0}\mathfrak{n}^{-1} \tag{2.68}$$

where to get the last equality we use (2.21) and (2.26). The expressions (2.67) and (2.68) describe a single deformed particle. We will discuss the properties of the Lagrangian (2.67) in the next two sections and will turn to the multi-particles system in Sect. 2.5.

Before finishing this section, let us rewrite the Lagrangian (2.67) in a slightly different way, which is going to be convenient for our investigations in Sect. 2.5 and reveals an important property of the momentum Π. To this end, let us revisit the continuity condition (2.64), which at $\phi = 0$ takes the form

$$\mathfrak{l}(0) = \mathfrak{n}\,\bar{\mathfrak{l}}(0)\,. \tag{2.69}$$

It is customary to partially fix the gauge so as to make the left hand side of this equation equal to the unit group element (or, equivalently, to absorb $\mathfrak{l}^{-1}(0)$ into the redefined $\mathfrak{n}$.) Then

$$\mathfrak{n} = \bar{\mathfrak{l}}^{-1}(0)\,, \tag{2.70}$$

so that

$$\Pi = \bar{\mathfrak{l}}^{-1}(0)\exp\left(\frac{m}{\kappa}J_0\right)\bar{\mathfrak{l}}(0)\,. \tag{2.71}$$

This equation has a very interesting interpretation. Namely recall that the Lorentz part of the connection is $A_i^{\mathcal{H}} = \mathfrak{l}^{-1}\partial_i\mathfrak{l}$, and therefore, the holonomy of $A^{\mathcal{H}}$ along the boundary Γ (see, e.g., [14]) equals

$$\mathrm{Hol}_{A^{\mathcal{H}}} \equiv \mathfrak{l}(0)\,\mathfrak{l}^{-1}(2\pi) = \bar{\mathfrak{l}}^{-1}(0)\exp\left(\frac{2\pi}{k}\mathscr{C}_J\right)\bar{\mathfrak{l}}(0) = \Pi\,, \tag{2.72}$$

where we used the fact that $\mathfrak{l}(0) = \mathbb{1}$, the fact that $\bar{\mathfrak{l}}(0) = \bar{\mathfrak{l}}(2\pi)$, and Eq. (2.64). Therefore, the group valued momentum Π has the geometric interpretation of the holonomy of connection $A^{\mathcal{H}}$, being the gravitational field, calculated along a curve surrounding the particle. As we will see the identification of the particle's momentum with the holonomy extends naturally to the multiparticle case and forces the momentum composition to be defined in terms of group multiplication.

2.4   Properties of the Deformed Lagrangian

Let us now turn to the detailed discussion of the effective particle Lagrangian (2.67). The first thing to notice is that, contrary to the free particle Lagrangian in (2.43), the group elements Π are nonlinear functions of momenta. This can be easily seen from the identity

$$\Pi = \mathfrak{n}e^{-\frac{m}{\kappa}J_0}\mathfrak{n}^{-1} = \exp\left(-\frac{m}{\kappa}\,\mathfrak{n}\,J_0\,\mathfrak{n}^{-1}\right) = \exp\left(-\frac{1}{\kappa}\,p_a\,J^a\right)\,. \tag{2.73}$$

Thus, the components of the momenta p_a are parameters defining the group element Π. It should be stressed that the choice of the group element parametrization is not unique and, in a sense, is equivalent to choosing a particular coordinate system on the group manifold. Since we are free to choose any such parametrization, it follows that the physical predictions of the theory could not depend on it: the theory should be momentum space coordinate-invariant.

The argument goes as follows. Consider the momentum measurements taken at CERN. There are computers that analyze the readings of the detectors and then do a complex calculation to finally provide physicists with the numerical values of the momenta of the particles. Suppose now that a mad saboteur hacks into the CERN system and changes the computer code so that the output now, instead of the components of momenta, will be some arbitrary functions of them $p_a \rightarrow \tilde{p}_a = f_a(p)$. Will the physics change? Of course not! The physics will be exactly the same (because nothing changes in the particles' reactions). Naively, therefore, we may conclude that the 'change of coordinates in momentum space' is physically irrelevant. There are, however, some condition that must be satisfied to guarantee that this change of coordinates, sometimes called the *change of basis* is consistently implemented. First, for dimensional reasons, the new momenta $\tilde{p}_a = f_a(p)$ should have the correct dimension. If the function $f_a(p)$ is nonlinear one has to have a universal momentum scale κ at the disposal to construct this function. Therefore, such construction will be highly unnatural in any theory that does not possess a 'built in' momentum scale, like, e.g., special relativity. By the same token, however, if such scale is present, it would be very odd if the general coordinate transformations in momentum space (with the fixed point being the zero momentum $p_a = 0$) is not a symmetry of the theory. Thus, in our case, where the momentum is defined by the group element Π, physics should not depend on particular coordinates one uses on the group manifold. There is yet another consistency condition that one has to employ, which guarantees that the momentum conservation rule transforms in an appropriate way, when the momentum space coordinates change. We will return to this issue later.

The group element Π can be then parametrized in two ways[8]

$$\Pi = \exp\left(-\frac{1}{\kappa} \, p_a \, J^a\right) = \mathbb{1} \, P_3 - \frac{1}{\kappa} \, P_a \, J^a \,, \tag{2.74}$$

where we included $1/\kappa$ factor to ensure that P_a have the correct dimension of momentum (Π is dimensionless by definition). Recalling that (2.37)

$$J_a \, J_b = \frac{1}{4} \, \eta_{ab} \, \mathbb{1} - \frac{1}{2} \, \epsilon_{ab}{}^c \, J_c \,,$$

we compute

$$P_3 = \cos\left(\frac{|p|}{2\kappa}\right) \,, \quad P_a = p_a \, \frac{2\kappa \sin\left(\frac{|p|}{2\kappa}\right)}{|p|} \,, \quad |p| = \sqrt{|p_a p^a|} \,. \tag{2.75}$$

By definition of the group the determinant of the matrix Π (2.74) must be equal one, and therefore

$$P_3^2 + \frac{1}{4\kappa^2} \left(P_0^2 - P_1^2 - P_2^2\right) = 1 \,. \tag{2.76}$$

---

[8]Some other useful parametrizations can be found in [15].

Therefore, the momentum space of our deformed particle is a three-dimensional Anti-de Sitter space.

Further, as in the discussion of the free particle in the preceding Sect. 2.2 discussion of the free particle, the coordinates p_a are not independent, but are constrained to belong to the orbit of the Lorentz group, $p_a J^a = mn J_0 n^{-1}$, and therefore, we have the mass-shell constraint

$$p_0^2 - p_1^2 - p_2^2 = m^2 , \tag{2.77}$$

which in terms of the P_a variables takes the form

$$P_0^2 - P_1^2 - P_2^2 = 4\kappa^2 \sin^2 \left(\frac{m}{2\kappa} \right) . \tag{2.78}$$

After these preparations, we can write the spinless part of the Lagrangian (2.67) in components

$$L = -\dot{P}_a P_3 x^a + \dot{P}_3 P_a x^a - \frac{1}{2\kappa} \epsilon^{bc}{}_a \dot{P}_b P_c x^a + N \left(P_a P^a - 4\kappa^2 \sin^2 \left(\frac{m}{2\kappa} \right) \right) , \tag{2.79}$$

where, as before, N is a lagrange multiplier enforcing the mass shell constraint (2.78) and

$$P_3 = \sqrt{1 + \frac{1}{4\kappa^2} P_a P^a} . \tag{2.80}$$

In the following discussion, it is convenient to use a slightly abbreviated, but much more illuminating form of the Lagrangian (2.79), to wit

$$L = -\dot{P}_\alpha E_a^\alpha(P) x^a + N \left(P_a P^a - 4\kappa^2 \sin^2 \left(\frac{m}{2\kappa} \right) \right) . \tag{2.81}$$

Notice that in (2.81), we start using Greek indices $\alpha, \beta, \dots$ to label the momentum variables. We do that to make it manifest that P_α are coordinates on curved momentum space manifold.

The geometry of momentum space is encapsulated in the form of the momentum frame field $E_a^\alpha(P)$, which in our case has the form

$$E_a^\alpha(P) = P_3 \delta_a^\alpha - \frac{1}{4\kappa^2} \frac{P^\alpha P_\beta}{P_3} \delta_a^\beta + \frac{1}{2\kappa} \epsilon^{\alpha\beta}{}_a P_\beta . \tag{2.82}$$

One easily checks that in the limit $\kappa \to \infty$, which corresponds to 'switching off' the gravitational interactions $E_a^\alpha(P) \to \delta_a^\alpha$ and the Lagrangian (2.81) becomes a Lagrangian of the standard, undeformed relativistic particle.

A remarkable thing about the deformed particle described by the Lagrangian (2.81) is that the Poisson bracket of positions does not vanish, and therefore, the corresponding quantum theory will be defined on non-commutative spacetime. Although spacetime non-commutativity refers, strictly speaking, to the quantum theory, we will use this term to describe the structure of the classical theory as well, in which case 'non-commutative' means 'non-vanishing Poisson bracket'.

It can be easily seen that the spacetime non-commutativity is directly related to the fact that the momentum space geometry is non-trivial. Indeed, let us consider the auxiliary variable $X^\alpha = E^\alpha_a(P) x^a$. In terms of these new variables, the Lagrangian (2.81) becomes that of an ordinary particle and thus the Poisson brackets for the phase space coordinates X^α, P_α are the standard ones

$$\{X^\alpha, P_\beta\} = \delta^\alpha_\beta, \quad \{X^\alpha, X^\beta\} = \{P_\alpha, P_\beta\} = 0. \tag{2.83}$$

Using $x^a = E^a_\alpha X^\alpha$, where E^a_α is the inverse frame field we get

$$\begin{aligned}
\{x^a, x^b\} &= \left(E^a_\alpha E^b{}_{\beta,}{}^\alpha E^\beta_c - a \leftrightarrow b\right) x^c \\
\{x^a, P_\alpha\} &= E^a_\alpha \\
\{P_\alpha, P_\beta\} &= 0,
\end{aligned} \tag{2.84}$$

and using (2.82), we find the following non-trivial Poisson bracket for the particle's coordinates:

$$\{x^a, x^b\} = \frac{1}{\kappa} \epsilon^{ab}{}_c x^c. \tag{2.85}$$

After quantization, this Poisson bracket will become the commutator, which tells us that quantum spacetime is non-commutative. As we will discuss in detail below, the emergence of spacetime non-commutativity is a direct consequence of the non-trivial geometry of momentum space.

The Lagrangian (2.81) is manifestly invariant (up to a local derivative) under global Lorentz transformations; it was actually constructed in such a way that Lorentz symmetry was manifest in all the steps of derivation. As we will see in the next section this is due to the properties we are using here: when different coordinates on momentum space are employed, the Lorentz symmetry may become deformed.

The Lagrangian (2.81) is clearly not invariant under the standard, global translations $\delta x^a = n^a$. It turns out, however that it is invariant (up to a total derivative) under deformed, momentum dependent translations

$$\delta x^a = E^a_\alpha(P) n^\alpha, \quad \delta P_\alpha = 0. \tag{2.86}$$

Thus, the translation is now momentum dependent, and this means that if two particles of different momenta move along the same worldline for one observer, their worldlines will not coincide with another observer, translated with respect to the original one. This is the first instance that the effect of *relative locality* exhibits itself.

2.5 The Case of Many Particles

In this section, we will generalize the construction presented in the Sect. 2.3 to the case of many particles. As we will see, the multiparticle Lagrangian contains not

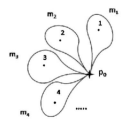

Fig. 2.2 The decomposition of the manifold in the multiparticle case

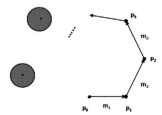

Fig. 2.3 The decomposition of the manifold in the multiparticle case

only the sum of the particles Lagrangians derived above, but also the terms that can be interpreted as describing topological interactions between particles.

Let us start by generalizing appropriately the decomposition of the spacial manifold that we made use of in Sect. 2.3. Suppose there are N particles. We take an arbitrary point p_0 and draw smooth, not intersecting, closed curves, starting and ending at p_0 such that the first curve, Γ_1, goes around the first particle, the second curve, Γ_2, goes around the second particle, etc, see Fig. 2.2.

It is worth noticing that each region containing a particle, whose boundary is Γ_n, $n = 1, \ldots, n$, has the topology of a punctured disc, just like the region $\mathscr{D}$ in Sect. 2.3. Therefore, we denote these regions $\mathscr{D}_n$. Let us now cut the regions $\mathscr{D}_n$ off the manifold. As a result, we get a manifold with n discs removed, with the boundaries of all the holes having a common point p_0. Now let us make a cut at p_0. As a result, we get a manifold, which in accordance with the notation above we call $\mathscr{H}$ with a single hole in it, whose inner boundary is a sum of the boundaries Γ_n, $\partial \mathscr{H} = \bigcup_n \Gamma_n$, see Fig. 2.3.

Exactly as it was in the case of a single particle we impose the continuity condition along each boundary Γ_n

$$\gamma_n|_{\Gamma_n} = \eta_n(t) \exp\left(\frac{1}{k}\mathscr{C}_n\phi_n\right)\bar{\gamma}_n\bigg|_{\Gamma_n}, \tag{2.87}$$

where each of the angular variables ϕ_n has the range $[0, 2\pi)$, as usual. A moment of reflection leads to the conclusion that all the steps of the derivation of the Lagrangian

(2.66) hold for each region bounded by Γ_n, and therefore, for many particles, we have

$$L = \frac{k}{2\pi} \sum_n \left\langle e^{-\frac{2\pi}{k}\mathscr{C}_{Jn}} \dot{\mathfrak{n}}_n^{-1} \mathfrak{n}_n e^{\frac{2\pi}{k}\mathscr{C}_{Jn}} \bar{\mathfrak{q}}_n - \dot{\mathfrak{n}}_n^{-1} \mathfrak{n}_n \, \bar{\mathfrak{q}}_n - \frac{2\pi}{k}\, \mathfrak{n}_n^{-1} \dot{\mathfrak{n}}_n \, \mathscr{C}_{Pn} \right\rangle. \quad (2.88)$$

Now we must take into account the crucial fact that the gauge group element γ_n that defines $A_n^{\mathscr{H}} = \gamma_n^{-1} \partial_i \gamma_n$ must be continuous at the ends of the segments Γ_n. Assuming that $\gamma_1(0) = \mathbb{1}$ we obtain a sequence of continuity conditions at the points $p_0, p_1, \ldots$. In order to evaluate (2.88), we need only the Lorentzian parts of these conditions, which read

$$\mathfrak{n}_1 \bar{\mathfrak{l}}_1 = \mathbb{1}$$

$$\mathfrak{n}_2 \bar{\mathfrak{l}}_2 = \bar{\mathfrak{l}}_1^{-1} \exp\left(\frac{2\pi}{k}\mathscr{C}_{J1}\right) \bar{\mathfrak{l}}_1 = \Pi_1^{-1}$$

$$\mathfrak{n}_3 \bar{\mathfrak{l}}_3 = \bar{\mathfrak{l}}_1^{-1} \exp\left(\frac{2\pi}{k}\mathscr{C}_{J1}\right) \bar{\mathfrak{l}}_1 \bar{\mathfrak{l}}_2^{-1} \exp\left(\frac{2\pi}{k}\mathscr{C}_{J2}\right) \bar{\mathfrak{l}}_2 = \Pi_1^{-1}\Pi_2^{-1}$$

$$\ldots = \ldots \qquad\qquad\qquad\qquad (2.89)$$

where it is understood that $\bar{\mathfrak{l}}_n$ denotes $\bar{\mathfrak{l}}_n(0)$. Notice that the first equation above is exactly the same as the one considered in the case of one particle, at the end of Sect. 2.3. Moreover, as it was in the one particle case, the expressions $\mathrm{Hol}_n = \bar{\mathfrak{l}}_n^{-1} \exp\left(\frac{2\pi}{k}\mathscr{C}_{Jn}\right) \bar{\mathfrak{l}}_n$ are nothing but the holonomies of the connection $A_n^{\mathscr{H}}$ along the n-th segment of the boundary.

Now we solve Eqs. (2.89) for $\mathfrak{n}_n$ and substitute it to (2.88). The resulting equations are getting rather complicated in general, so here we will write down explicitly and discuss only expressions pertaining to the two particles case. In this case, the Lagrangian has the form

$$L = \kappa \left\langle \dot{\Pi}_1 \, \Pi_1^{-1} \, \mathfrak{x}_1 \right\rangle + \kappa \left\langle \dot{\Pi}_2 \, \Pi_2^{-1} \, \mathfrak{x}_2 \right\rangle + \kappa \left\langle \left(\Pi_2 \, \dot{\Pi}_1 \, \Pi_1^{-1} \, \Pi_2^{-1} - \dot{\Pi}_1 \, \Pi_1^{-1} \right) \mathfrak{x}_2 \right\rangle$$
$$+ \left\langle \dot{\bar{\mathfrak{l}}}_1 \, \bar{\mathfrak{l}}_1^{-1} \, \mathscr{C}_{P1} \right\rangle + \left\langle \dot{\bar{\mathfrak{l}}}_2 \, \bar{\mathfrak{l}}_2^{-1} \, \mathscr{C}_{P2} \right\rangle + \left\langle \bar{\mathfrak{l}}_2 \, \dot{\Pi}_1 \, \Pi_1^{-1} \bar{\mathfrak{l}}_2^{-1} \, \mathscr{C}_{P2} \right\rangle. \quad (2.90)$$

Let us discuss this Lagrangian, omitting the spin terms for a moment. To understand the physical meaning of (2.86), it is convenient to use the 'centre' position[9] $\mathfrak{x} = \frac{1}{2}(\mathfrak{x}_1 + \mathfrak{x}_2)$ and the relative position $\mathfrak{d} = \frac{1}{2}(\mathfrak{x}_1 - \mathfrak{x}_2)$, in terms of which the spinless part of the Lagrangian (2.90) takes the form

$$L = \kappa \left\langle \dot{\Pi} \, \Pi^{-1} \, \mathfrak{x} \right\rangle - \kappa \left\langle \dot{\Pi} \, \Pi^{-1} \, \mathfrak{d} \right\rangle + 2\kappa \left\langle \dot{\Pi}_1 \, \Pi_1^{-1} \, \mathfrak{d} \right\rangle, \quad (2.91)$$

where

$$\Pi \equiv \Pi_2 \Pi_1 \qquad\qquad\qquad (2.92)$$

---

[9]Note that this is *not* the centre of mass position.

has an interpretation of the total momentum of our two-particles system.

Notice now that the Lagrangian (2.91) is invariant under rigid translations, which do not change the relative distance between the particles $\delta\mathfrak{d} = 0$ or $\delta\mathfrak{x}_1 = \delta\mathfrak{x}_2$, such that (cf. (2.81), (2.82))

$$\delta\delta\mathfrak{x}^a = E_\alpha^a (P_1 \oplus P_2) \, n^\alpha \,, \tag{2.93}$$

where $(P_1 \oplus P_2)_\alpha$ are coordinates of the total momentum of the system. These coordinates can be calculated from Eq. (2.92)

$$\Pi(P_1 \oplus P_2) = \Pi_2(P_2) \, \Pi_1(P_1) \,, \tag{2.94}$$

which gives for $\Pi_n = P_3^{(n)} \, \mathbb{1} + \frac{1}{\kappa} \, P_\alpha^{(n)} \, J^\alpha$

$$(P_1 \oplus P_2)_\alpha = P_3^{(1)} \, P_\alpha^{(2)} + P_3^{(2)} \, P_\alpha^{(1)} + \frac{1}{\kappa} \, \epsilon_\alpha{}^{\beta\gamma} \, P_\beta^{(2)} P_\gamma^{(1)} \,. \tag{2.95}$$

It follows immediately[10] from (2.93) and (2.95) that the components of the total conserved momentum are indeed $(P_1 \oplus P_2)_\alpha$. Therefore, (2.95) can be understood as a deformed momenta composition rule.

It is would be quite illuminating to compare this with the case in which the Lagrangian of two particles contains only the first two terms of (2.91). In that case, the global translational symmetry transformations have the form

$$\delta\mathfrak{x}_1^a = E_\alpha^a(P_1) \, n^\alpha \,, \quad \delta\mathfrak{x}_2^a = E_\alpha^a(P_2) \, n^\alpha \,, \tag{2.96}$$

and the total conserved momentum is the usual sum of momenta $P_\alpha = P_{1\alpha} + P_{2\alpha}$. Obviously, in this case, the translation does not preserve the relative position of the particles $\delta\mathfrak{d} \neq 0$ and the theory with such form of the Lagrangian exhibit the *relative locality* phenomenon, i.e., events that are local for one observer, for example, crossing of particle's worldlines at a point are not local for the translated one [16, 17].

2.6 κ-Carrollian Deformation

In the previous section, we discussed a particular deformation of particle dynamics emerging from 2+1 gravity. It is natural to ask if this deformation is unique and it turns out that the answer is negative. There exist at least one different deformed particle dynamics that could be derived in the case of vanishing cosmological constant, and there are some indications that there might be more.

---

[10]To see this it is sufficient to recall that the conserved Noether current associated with a global symmetry with parameters n^α can be most easily found by taking n^α time dependent and looking for the term in variation of the Lagrangian that is proportional to the time derivative on n^α (the terms proportional to n^α itself vanish because this parameter generates global symmetry.).

Let us start with describing the new deformation that we will call κ-Carrollian [18]. Consider the spacetime symmetry algebra (2.2) in the de Sitter case $\Lambda > 0$. If we take the contraction limit $\Lambda \to 0$ of this algebra, we get as a result the standard Poincaré algebra, which produces, as it was explained above, the deformed particle dynamics with curved SO(2, 1) momentum space. But this is not the only contraction of the $\mathfrak{so}(3, 1)$ algebra that one can think of. In fact, there is another contraction which leads to the deformed particle dynamics with different momentum space.

To construct this contraction, let us rewrite the algebra (2.2) in a slightly different, but equivalent, form, absorbing the cosmological constant into the generators P, so that all the generators are dimensionless now

$$[J_a, J_b] = \epsilon_{ab}{}^c J_c , \quad [J_a, P_b] = \epsilon_{ab}{}^c P_c , \quad [P_a, P_b] = -\epsilon_{ab}{}^c J_c . \tag{2.97}$$

In the first step of our construction, we introduce new generators S_a, such that

$$S_a = P_a + \epsilon_{a0b} J^b , \tag{2.98}$$

so that, we have

$$[J_a, J_b] = \epsilon_{ab}^{\ c} J_c , \quad [J_a, S_b] = \epsilon_{ab}^{\ c} S_c - \eta_{ab} J_0 + \eta_{b0} J_a , \quad [S_a, S_b] = \eta_{a0} S_b - \eta_{b0} S_a . \tag{2.99}$$

It should be stressed at this point that the algebra (2.99) is still equivalent to the original one (2.97) since the only thing we did was to rewrite the latter in terms of new generators that are linear combination of the old ones.

Now we are ready to make a decisive step. Let us now rescale $\tilde{J}_a \equiv \sqrt{\Lambda} J_a$ to obtain

$$[\tilde{J}_a, \tilde{J}_b] = \sqrt{\Lambda} \, \epsilon_{abc} \tilde{J}^c ,$$
$$[\tilde{J}_a, S_b] = \sqrt{\Lambda} \, \epsilon_{abc} S^c + (\eta_{b0} \tilde{J}_a - \eta_{ab} \tilde{J}_0) ,$$
$$[S_a, S_b] = \eta_{a0} S_b - \eta_{b0} S_a , \tag{2.100}$$

which after contraction $\Lambda \to 0$ takes the form

$$[\tilde{J}_a, \tilde{J}_b] = 0 , \quad [\tilde{J}_a, S_b] = (\eta_{b0} \tilde{J}_a - \eta_{ab} \tilde{J}_0) , \quad [S_a, S_b] = \eta_{a0} S_b - \eta_{b0} S_a . \tag{2.101}$$

The Lie algebra generated by the generators S_a is called an(2) algebra since it contains one abelian generator S_0 and two nilpotent ones S_i $i = 1, 2$. We will postpone the detailed discussion of this algebra and the corresponding group AN(2) to Chaps. 4 and 5, but it should be mentioned here that the group manifold of AN(2) is a half of de Sitter space. This manifold is well known to cosmologists because it is the part of de Sitter space covered by the widely used (e.g., in the context of the inflationary cosmological model) flat cosmological coordinates.

In terms of the new generators, the scalar products read

$$\langle \tilde{J}_a S_b \rangle = \sqrt{\Lambda} \, \eta_{ab} , \quad \langle \tilde{J}_a \tilde{J}_b \rangle = \langle S_a S_b \rangle = 0 . \tag{2.102}$$

In spite of the fact that this scalar product becomes degenerate in the limit $\Lambda \to 0$, in the effective particle action Λ cancels out as a result of the presence of $(\Lambda)^{-1/2}$ factor in the prefactor $k/2\pi$, see (2.22), and the contraction limit is not singular.

A gauge group element can be decomposed into a product of an element $\mathfrak{j}$ obtained exponentiating the abelian generators $\tilde{J}_a$ and an element $\mathfrak{s}$ of the group $\mathrm{an}(2)$

$$\gamma = \mathfrak{j}\,\mathfrak{s} = (1 + \iota^a\,\tilde{J}_a)\,e^{\sigma^i\,S_i}\,e^{\sigma^0\,S_0}, \quad i = 1,2, \tag{2.103}$$

where for $\mathfrak{s}$ we use the parametrization that proved convenient in the context κ-Poincaré theories (see Chap. 4) and is related to another parametrization $\mathfrak{s} = \xi_3 + \xi^a S_a$ via $\sigma^0 = 2\log(\xi_3 + \frac{1}{2}\xi^0)$, $\sigma^i = (\xi_3 + \frac{1}{2}\xi^0)\,\xi^i$.

We must also change the form of $\mathscr{C} = \mathscr{C}_J + \mathscr{C}_S$, describing the particle at rest, so as to have the mass in the S sector. Adjusting dimensions properly, we get $\mathscr{C}_J = s/\sqrt{\Lambda}\,\tilde{J}_0$, $\mathscr{C}_S = m/\sqrt{\Lambda}\,S_0$.

After these preparatory steps, we can turn to the derivation of the deformed particle action. Revisiting the continuity condition (2.61) and writing $\eta = (1 + n)\,\mathfrak{h}$, with $\mathfrak{h} \in \mathrm{AN}(2)$, $n = n^a\,J_a$ and using the factorization (2.103) we find that

$$\mathfrak{s}^{-1} = \mathfrak{h}\exp\left(\frac{1}{k}\mathscr{C}_S\phi\right)\bar{\mathfrak{s}}^{-1}. \tag{2.104}$$

Next, from the commutation relations (2.101), for an arbitrary $\mathfrak{s}$, we have

$$\mathfrak{s}\exp\left(\frac{1}{k}\mathscr{C}_J\phi\right)\mathfrak{s}^{-1} = \exp\left(\frac{1}{k}\mathscr{C}_J\phi\right). \tag{2.105}$$

As a result, we obtain the second condition

$$\mathfrak{u} = \exp\left(\frac{1}{k}\mathscr{C}_J\phi\right)\mathfrak{s}\,(1 - n)\mathfrak{s}^{-1}, \tag{2.106}$$

where we denote $\mathfrak{u} \equiv \bar{\mathfrak{j}}^{-1}\mathfrak{j}$.

Plugging all these ingredients into the action (2.51) and repeating almost verbatim all the steps of Sect. 2.3, we obtain an expression very similar to (2.67), namely

$$L = \frac{k}{2\pi}\langle\Pi\,\dot{\Pi}^{-1}\,x\rangle + \langle\mathscr{C}_J\bar{\mathfrak{s}}^{-1}\dot{\mathfrak{s}}\rangle, \tag{2.107}$$

but with the deformed momenta Π being now the elements of the $\mathrm{AN}(2)$ group, instead of the Lorentz group $\mathrm{SO}(2, 1)$, to wit

$$\Pi \equiv \bar{\mathfrak{s}}\,e^{\frac{2\pi}{k}\mathscr{C}_S}\bar{\mathfrak{s}}^{-1}, \quad x \equiv \bar{\mathfrak{s}}\,\mathfrak{h}^{-1}(0)n(0)\mathfrak{h}(0)\bar{\mathfrak{s}}^{-1}. \tag{2.108}$$

Since $\mathrm{AN}(2)$ is a submanifold of the three-dimensional de Sitter space, we managed to obtain the momentum space of positive, instead of the negative, constant curvature. Moreover, contrary to (2.67), which is defined only in 2+1 dimension (because the dimension of the Lorentz group equals the spacetime dimensions only in the case of $2 + 1$-dimensional spacetime), the expression (2.108) can be readily generalized to any spacetime dimension.

In the following discussion, we will consider only the spinless case. It turns out that this is in fact the most general case, because the spin term does not contribute non-trivially to the equations of motion. Indeed, an arbitrary variation $\delta \bar{s} = \varpi \, \bar{s}$, $\delta \bar{s}^{-1} = -\bar{s}^{-1} \varpi$ of the spin term in (2.108) results in a total time derivative

$$\delta \left(\mathscr{C}_J \bar{s}^{-1} \dot{\bar{s}} \right) = \left(\bar{s} \mathscr{C}_J \bar{s}^{-1} \dot{\varpi} \right) = \frac{d}{dt} \left\langle \mathscr{C}_J \varpi \right\rangle , \tag{2.109}$$

because $\tilde{J}_0$ commutes with all S generators (cf. (2.101).)

Let us now turn to the detailed discussion of the properties of Lagrangian (2.107). The first thing to notice is that the definition of the group valued momentum Π in (2.108) puts severe restrictions on its form. Indeed, if we write

$$\Pi = e^{p^i / \kappa \, S_i} \, e^{p^0 / \kappa \, S_0} , \tag{2.110}$$

and take $\bar{s}$ to have the form

$$\bar{s} = e^{\bar{\sigma}^i \, S_i} \, e^{\bar{\sigma}^0 \, S_0} , \tag{2.111}$$

we immediately find that

$$p^0 = m , \qquad p^i = \kappa \, (1 - e^{\frac{m}{\kappa}}) \, \bar{\sigma}^i . \tag{2.112}$$

As it was in the case considered above, these equations play a role of the mass-shell relation, and force the energy to be constant, independently of the particle dynamics.[11] In the undeformed case (which can be obtained in the limit $\kappa \to \infty$), such mass-shell condition makes the particle effectively frozen, it can not move, and for that reason, following [19, 20] we call it the 'Carroll particle', since it is a nice example of physics of the reign of Red Queen:

"Well, in our country," said Alice, still panting a little, "you'd generally get to somewhere else if you run very fast for a long time, as we've been doing."

"A slow sort of country!" said the Queen. "Now, here, you see, it takes all the running you can do, to keep in the same place. If you want to get somewhere else, you must run at least twice as fast as that!" [21]

---

[11] Although the spinless part of the Lagrangian (2.107) looks exactly the same the Lagrangian of the κ-deformed particle that we are going to discuss in detail in Chap. 4, (4.142) there is a crucial difference between the two. In the case of κ-deformation, we do not impose the constraint (2.108), which makes it possible for κ-deformed particle to be not frozen.

Using the first equation in (2.112) as a mass-shell condition and the expression for Π (2.110), we can rewrite the Lagrangian (2.107) in components

$$L = x^0 \dot{p}_0 + x^i \dot{p}_i - \kappa^{-1} x^i p_i \dot{p}_0 + \lambda(p_0^2 - m^2), \qquad (2.113)$$

where, as before, we introduce the Lagrange multiplier λ to enforce the mass-shell constraint. The equations of motion following from variations over x are momentum conservations $\dot{p}_a = 0$, while the ones resulting from the variation over momenta give

$$\dot{x}^0 = 2\lambda p_0 = 2\lambda m, \quad \dot{x}^i = 0, \qquad (2.114)$$

so that indeed the Carroll particle is always at rest. Furthermore, from (2.108) we find the explicit expressions for the components $x^0 = n^0 - (e^{\zeta_0 - \bar{\sigma}_0} \bar{\sigma}^i - \zeta^i) n_i$, $x^i = e^{\zeta_0 - \bar{\sigma}_0} n^i$. Then (2.114) provides us with conditions for coordinates of n, $\mathfrak{h} = e^{\zeta^i S_i} e^{\zeta^0 S_0}$ and $\bar{\mathfrak{s}}$.

The presence of the nonlinear, deformed term in the Lagrangian (2.113) results in the non-trivial Poisson bracket algebra of the κ-deformed phase space [22]

$$\left\{ x^i, p_j \right\} = \delta^i_j, \quad \left\{ x^0, p_0 \right\} = 1, \quad \left\{ x^0, p_i \right\} = -\frac{1}{\kappa} p_i, \quad \left\{ x^0, x^i \right\} = \frac{1}{\kappa} x^i, \qquad (2.115)$$

We will discuss this phase space in more detail in Chap. 4.

The symmetries of the action obtained from the Lagrangian (2.114) form the algebra of infinitesimal deformed Carroll transformations, consisting of

- rotations

$$\delta x^i = \rho \, \epsilon^i{}_j x^j, \quad \delta p_i = \rho \, \epsilon_i{}^j p_j, \quad \delta x^0 = \delta p_0 = 0; \qquad (2.116)$$

- deformed boosts

$$\delta x^0 = (1 + \kappa^{-1} p_0) \lambda_i x^i, \quad \delta p_i = -\lambda_i p_0, \quad \delta x^i = \delta p_0 = 0; \qquad (2.117)$$

- deformed translations

$$\delta x^0 = a^0, \quad \delta x^i = e^{p_0/\kappa} a^i, \quad \delta p_a = 0; \qquad (2.118)$$

- the spatial conformal transformation

$$\delta x^i = \eta \, x^i, \quad \delta p_i = -\eta \, p_i, \quad \delta x^0 = \delta p_0 = 0, \qquad (2.119)$$

where ρ, λ_i, a^a, η are parameters of the respective transformations.

Actually, as noted in [20] the undeformed Carroll particle has an infinite-dimensional symmetry. This property holds in the deformed case as well, and the generator of the infinitesimal symmetry transformations $\delta\phi^a = \{\phi^a, G\}$, where ϕ is an arbitrary function on phase space, is given by

$$G = f(p_0/\kappa) p_0 \, \xi^0(x^i) + p_i \, \xi^i(x^i), \quad f(0) = 1, \tag{2.120}$$

where $f(p_0/\kappa)$ is an arbitrary function of energy, while $\xi^i(x^i)$, $\xi^0(x^i)$ are arbitrary functions of position.

2.7   Deformed Spacetime Symmetries of 2+1 Quantum Gravity

In the preceding sections, we noticed that the deformations arise as a result of incorporating the would be gauge degrees of freedom of gravity into the particles' dynamics. This effect is, as we saw, purely classical in 2+1 dimensions, as a result of the fact that the Newton's constant has the dimension of inverse mass and the Planck mass does not have $\hbar$ in its definition. On the other hand, the Planck length $\ell_{PL} = \hbar G$ requires $\hbar$, and therefore, is a scale relevant for the quantum structure of spacetime. Let us note in passing that neglecting quantum effects is equivalent to entering the relative locality regime [16,17]. It is not clear, therefore, what is the relation between the deformed action of a particle and (possibly) deformed symmetries of quantum spacetime. In this section, based on [23,24], we will attempt to answer this question.

Quantum gravity changes dramatically the spacetime structure at small distances, allowing for fluctuations of spacetime itself. As a result, the symmetries of quantum spacetime, even in the flat limit, where spacetime curvature is negligible, should differ from the classical Poincaré symmetries. Over many years, more and more circumstantial evidence have accumulated indicating that the symmetries of flat quantum spacetime should be described by some kind of quantum deformation of the Poincaré group, and that the infinitesimal generators of the deformed group are to be described by a Hopf algebra, which is a deformation of the Poicaré algebra.

In what follows, we will be interested in *symmetries flat quantum spacetime* and we start explaining the meaning of the phrase. In classical general relativity, any spacetime can be identified with some particular configuration of the gravitational field. In particular, the flat Minkowski spacetime is the configuration characterized by Minkowski metric $\eta_{\mu\nu} = (-1, 1, 1, \ldots)$ and the maximal sets of its Killing vectors

$$\left\{ \partial_\mu \, , x_\mu \partial_\nu - x_\nu \partial_\mu \right\} \tag{2.121}$$

generating translations and Lorentz transformations, satisfying the standard Poincaré algebra. This same algebra can be also derived from dynamics of the gravitational field. Since gravity is a gauge theory with reparametrization invariance, i.e., a theory with vanishing Hamiltonian, its dynamics is completely characterized[12] by the system of constraints and the algebra that the constraints satisfy. In the case of Einstein

---

[12]In the case when spacetime has no boundaries. When boundaries are present the situation is by far more complicated and interesting. See [25] for details.

gravity (in 3D), we have to do with three constraints: two diffeomorphism constraints $\mathscr{D}[\mathbf{f}]$ generating a spacial diffeomorphism and the Hamiltonian constraint $\mathscr{H}[g]$ generating time translation, with $\mathbf{f}$ and g being the appropriate smearing functions. The constraints $\mathscr{D}[\mathbf{f}]$, $\mathscr{H}[g]$ are built from the components of the metric (triad) and its canonical momenta see [3] for details, and in quantum gravity become quantum operators. They satisfy the Poisson bracket algebra

$$\{\mathscr{D}[\mathbf{f}_1], \mathscr{D}[\mathbf{f}_2]\} = \mathscr{D}[[\mathbf{f}_1, \mathbf{f}_2]] \tag{2.122}$$

$$\{\mathscr{D}[\mathbf{f}], \mathscr{H}[g]\} = \mathscr{H}[f^a \partial_a g] \tag{2.123}$$

$$\{\mathscr{H}[g_1], \mathscr{H}[g_2]\} = \mathscr{D}[\mathbf{f}(g_1, g_2)] \tag{2.124}$$

where

$$[\mathbf{f}_1, \mathbf{f}_2] = f_1^a \partial_a \mathbf{f}_2 - f_2^a \partial_a \mathbf{f}_1 \qquad f^a(g_1, g_2) = h^{ab}(g_1 \partial_b g_2 - g_2 \partial_b g_1), \tag{2.125}$$

and h^{ab} is the inverse spacial metric. It can be checked that if the smearing functions are the Killing vectors of the flat Minkowski space and the metric h^{ab} is the Minkowski space metric the algebra above becomes the Poincaré algebra. Let us see how it comes about.

First, we consider translations represented by constraints smeared by the constant functions. $g = a$, $\mathbf{f} = \mathbf{a}$, and we call them energy and momenta

$$E \equiv \mathscr{H}[a], \quad \mathbf{P} = \mathscr{D}[\mathbf{a}] \tag{2.126}$$

Then it follows from (2.125) that these three constraints commute. Let us then consider infinitesimal Lorentz transformations, consisting of infinitesimal rotations $g = 0$, $f^a = \rho^a{}_b x^b$ with ρ^{ab} being an antisymmetric matrix,[13] and infinitesimal boosts, represented by $g = -\xi \mathbf{x} = -\xi_a x^a$, $\mathbf{f} = 0$. They satisfy the following algebra:

$$\{\mathscr{D}[\rho_1^a{}_b x^b], \mathscr{D}[\rho_2^a{}_b x^b]\} = \mathscr{D}\left[(\rho_1^a{}_c \rho_2^c{}_b - \rho_2^a{}_c \rho_1^c{}_b)x^b\right]$$

$$\{\mathscr{D}[\rho^a{}_b x^b], \mathscr{D}[\mathbf{a}]\} = \mathscr{D}\left[\rho^a{}_b a^b\right]$$

$$\{\mathscr{D}[\rho^a{}_b x^b], \mathscr{H}[a]\} = 0 \tag{2.127}$$

so that the bracket of two rotations is the rotations, and the rotation acts on translations as it should (it does not act the time translation).

Let us now consider the boost action on translations

$$\{\mathscr{H}[-\xi \mathbf{x}], \mathscr{D}[\mathbf{a}]\} = \mathscr{H}[\xi \mathbf{a}]$$

$$\{\mathscr{H}[-\xi \mathbf{x}], \mathscr{H}[a]\} = \mathscr{D}[a\xi] \tag{2.128}$$

---

[13]In three dimensions there is only one rotation generator with $\rho^{ab} = \rho^{12}$ and its bracket with itself obviously vanishes. Equation (2.127) holds in any dimension.

so that the boost transforms the energy E into momentum $\mathbf{P}$ and vice versa. Finally

$$\left\{\mathcal{H}[-\xi_1\mathbf{x}],\ \mathcal{H}[-\xi_2\mathbf{x}],\right\} = -\mathcal{D}[(\xi_1^a\xi_{2b} - \xi_2^a\xi_{1b})x^b] \tag{2.129}$$

and the bracket of two boosts is proportional to rotations, as it should.

The same exercise can be repeated in the case of the three-dimensional Euclidean de Sitter space, with cosmological constant Λ and the resulting algebra of generators corresponding to energy E, linear momenta P_a, rotation M and boosts N_a has the form

$$\{E, P_a\} = \Lambda N_a, \qquad \{P_1, P_2\} = \Lambda M$$
$$\{N_a, E\} = P_a, \qquad \{N_a, P_b\} = -\delta_{ab}E, \qquad \{N_1, N_2\} = M, \tag{2.130}$$
$$\{M, N_a\} = \epsilon_{ab}N_b, \qquad \{M, P_a\} = \epsilon_{ab}P_b, \qquad \{M, E\} = 0.$$

In quantum theory, the Poisson brackets (2.122)–(2.124) are replaced by commutators and the smeared phase space constraints $\mathcal{D}[\mathbf{f}]$ and $\mathcal{H}[g]$ become quantum mechanical operators, whose commutators have to be computed.

Three-dimensional Euclidean[14] gravity with a positive cosmological constant is defined by a Chern-Simons action (2.16) with the gauge group $SO(4)$. After decomposing the connection A into connection ω and triad e we find

$$I = \frac{\kappa}{2}\int_{\mathcal{M}} d^3x\, \epsilon^{\mu\nu\rho}\left\langle e_\mu F_{\nu\rho}(\omega) + \frac{\Lambda}{3}e_\mu e_\nu e_\rho\right\rangle. \tag{2.131}$$

where $F_{\nu\rho}(\omega)$ is a curvature of connection ω. Notice that since we are considering here Euclidean gravity, both the connection and triad are valued in the algebra $\mathfrak{so}(3) \simeq \mathfrak{su}(2)$.

We canonically decompose the three-dimensional spacetime into a time direction and a surface Σ, namely $\mathcal{M} = \Sigma \times \mathbb{R}$. Then the canonical phase space is parametrized by the value of ω on Σ, which we denote $A_a^i = 1/2\epsilon^i{}_{jk}\omega_a^{jk}$, and its conjugate momentum $E_j^b = \kappa\epsilon^{bc}e_c^k\eta_{jk}$. In our notation, $a = 1, 2$ are space coordinate indices on Σ, $i, j = 1, 2, 3$ label internal $\mathfrak{su}(2)$ indices, which we raise and lower with the Killing metric δ_{ij}, and $\epsilon^{ab} = -\epsilon^{ba}$ with $\epsilon^{12} = 1$. The canonical phase space variables satisfy the Poisson bracket

$$\{A_a^i(x), E_j^b(y)\} = \delta_a^b\delta_j^i\delta^{(2)}(x, y). \tag{2.132}$$

---

[14] We use here Euclidean not Lorentzian signature gravity for technical reasons, which are explained in [23]. It is believed that the final result can be continued back to Lorentzian case (like it is in the case of Wick rotation).

The variation of the action with respect to Lagrange multipliers e_t^i and ω_t^{ij} leads to two sets of smeared constraints

$$G[\alpha] = \int_\Sigma \alpha^i G_i = \int_\Sigma \alpha^i D_A E_i = 0 \,, \tag{2.133}$$

$$C_\Lambda[N] = \int_\Sigma N_i C_\Lambda^i = \int_\Sigma N_i (\kappa F^i(A) + \frac{\Lambda}{2\kappa} \epsilon^{ijk} E_j \wedge E_k) = 0 \,, \tag{2.134}$$

where α, N are arbitrary su(2)-valued test functions, independent of the connection and momentum variables. The constraint (2.133) is called the Gauss constraint and it implements the local $SU(2)$ gauge invariance of the theory; the second constraint (2.134) is called the curvature constraint and it encodes the information that the connection is no longer flat (as it was in the $\Lambda = 0$ case) and it also generates gauge symmetries.

The classical constraint algebra of the theory reads

$$\begin{aligned}
\{C_\Lambda[N], C_\Lambda[M]\} &= \Lambda\, G[[N, M]] \,, \\
\{C_\Lambda[N], G[\alpha]\} &= C_\Lambda[[N, \alpha]] \,, \\
\{G[\alpha], G[\beta]\} &= G[[\alpha, \beta]] \,,
\end{aligned} \tag{2.135}$$

where $[a, b]^i = \epsilon^i{}_{jk} a^j b^k$ is the commutator of su(2). As shown in [23], the constraints $G[\alpha]$ and $C_\Lambda[N]$ are in one to one correspondence with the diffeomorphism and hamiltonian constraint $\mathscr{D}[\mathbf{f}]$, $\mathscr{H}[g]$ discussed above, and their algebra reduces to the Poincaré algebra in the flat space limit.

As it turns out, to quantize three-dimensional Euclidean de Sitter gravity, described briefly above, we need to define a new non-commutative connection [26,27]

$$\Omega_a^{\pm i} = A_a^i \pm \sqrt{\Lambda} e_a^i = A_a^i \pm \frac{\sqrt{\Lambda}}{\kappa} \epsilon_{ba} E_i^b \,, \tag{2.136}$$

for which the canonical Poisson bracket has the form

$$\{\Omega_a^{\pm i}(x), \Omega_b^{\pm j}(y)\} = \pm 2 \frac{\Lambda}{\kappa} \epsilon_{ab} \delta^i_j \delta^{(2)}(x, y) \,, \quad \{\Omega_a^{\pm i}(x), \Omega_b^{\mp j}(y)\} = 0 \tag{2.137}$$

In terms of this connection Gauss and curvature constraints can be expressed as

$$C_\Lambda[N] = \frac{1}{2} \left(H^+[N] + H^-[N] \right) \,, \tag{2.138}$$

$$G[N] = \frac{1}{2\sqrt{\Lambda}} \left(H^+[N] - H^-[N] \right) \,, \tag{2.139}$$

where

$$H^\pm[N] \equiv \kappa \int_\Sigma N_i F^i(\Omega^\pm) \tag{2.140}$$

is the curvature constraint for the non-commutative connection Ω.

The set of constraints (2.140) is equivalent to the constraints (2.133), (2.134), and their algebra is

$$\{H^{\pm}[N], H^{\pm}[M]\} = \pm 2\sqrt{\Lambda}\, H^{\pm}[[N, M]]$$
$$\{H^{+}[N], H^{-}[M]\} = 0, \tag{2.141}$$

corresponding to two copies of $su(2)$, i.e., the constraints of the theory generate a local $su(2) \oplus su(2)$ symmetry.

After quantization, the Poisson brackets become commutators and it turns out that the algebra of quantum constraints $\hat{H}^{\pm}[N]$ becomes *deformed* and instead of (2.141) takes the form [26]

$$\left[\hat{H}^{\pm}[N_p], \hat{H}^{\pm}[M_p]\right]|\Psi\rangle = \pm\sqrt{\Lambda}\left(q + q^{-1}\right)\hat{H}^{\pm}[[N_p, M_p]]|\Psi\rangle,$$
$$\left[\hat{H}^{\pm}[N_p], \hat{H}^{\mp}[M_p]\right]|\Psi\rangle = 0. \tag{2.142}$$

with the deformation parameter $q = \exp(i\hbar\sqrt{\Lambda}/2\kappa)$ and $|\Psi\rangle$ being an arbitrary state (not the physical one). In the classical limit $\hbar \to 0$, the quantum constraints algebra (2.142) becomes the classical one (2.141).

Returning to the physical generators and setting $\hbar = 1$ from (2.142), we obtain the deformed counterpart of the classical algebra (2.130)

$$[E, P_a] = \Lambda N_a, \qquad [N_a, E] = P_a,$$
$$[P_1, P_2] = \Lambda \frac{\sinh(zM)}{\sin(z)}\cosh\left(zE/\sqrt{\Lambda}\right),$$
$$[N_a, P_b] = -\delta_{ab}\sqrt{\Lambda}\frac{\sinh\left(zE/\sqrt{\Lambda}\right)}{\sin(z)}\cosh(zM), \tag{2.143}$$
$$[N_1, N_2] = \frac{\sinh(zM)}{\sin(z)}\cosh\left(zE/\sqrt{\Lambda}\right),$$
$$[M, N_a] = \epsilon_a{}^b N_b, \qquad [M, P_a] = \epsilon_a{}^b P_b, \qquad [M, E] = 0.$$

where the deformation parameter $z = \sqrt{\Lambda}/2\kappa$.

After rescaling the generators [30]

$$E \to E, \qquad M \to M, \qquad P_a \to e^{z\tilde{E}/\left(2\sqrt{\Lambda}\right)}\tilde{P}_a,$$
$$N_a \to e^{z\tilde{E}/\left(2\sqrt{\Lambda}\right)}\left(\tilde{N}_a - \frac{z}{2\sqrt{\Lambda}}\epsilon_{ab}\tilde{M}\tilde{P}_b\right), \tag{2.144}$$

we can now take the limit $\Lambda \to 0$, keeping κ constant and finite to obtained the deformed algebra of symmetries of flat quantum spacetime [28,29]

$$[E, P_a] = [P_1, P_2] = 0, \qquad [N_a, E] = P_a,$$

$$[N_a, P_b] = -\delta_{ab} \left(\frac{\kappa}{2} \left(1 - e^{-2E/\kappa} \right) - \frac{1}{2\kappa} \mathbf{P}^2 \right) - \frac{1}{\kappa} P_a P_b,$$

$$[N_1, N_2] = M, \qquad [M, N_a] = \epsilon_{ab} N_b, \qquad [M, P_a] = \epsilon_{ab} P_b, \qquad [M, E] = 0.$$
$$(2.145)$$

This is the celebrated κ-Poincaré algebra in three dimensions.

Let us finish this section with a few remarks. As it turns out [23], the algebra of spacetime symmetries is not the only object that becomes deformed. Importantly, the algebra of symmetries in the quantum case is deformed to a non-trivial Hopf algebra. It has a non-trivial r-matrix (see Sect. 4.1), coproduct, and antipode (see Sect. 5.1). These additional structures make the deformed spacetime symmetry algebra really 'rigid', such that its fundamental properties could not be changed by a mere change of basis (the algebra (2.145) can be brought to the form of the standard Poincaré algebra by a nonlinear change of basis, but the other structures, r-matrix, coproduct, and antipode remain non-trivial in this new basis). We will say more about these structures and Hopf algebras in the Part II below.

References

1. Witten, E.: (2+1)-dimensional gravity as an exactly soluble system. Nucl. Phys. B **311**, 46 (1988)
2. Achucarro, A., Townsend, P.K.: A Chern-Simons action for three-dimensional anti-De Sitter supergravity theories. Phys. Lett. B **180**, 89 (1986)
3. Carlip, S.: Quantum Gravity in 2+1 Dimensions, 276p. University Press, Cambridge, UK (1998)
4. Balachandran, A.P., Marmo, G., Skagerstam, B.S., Stern, A.: Gauge theories and fibre bundles - applications to particle dynamics. Lect. Notes Phys. **188**, 1–140 (1983). https://doi.org/10.1007/3-540-12724-0_1, arXiv:1702.08910 [quant-ph]
5. de Sousa Gerbert, P.: On spin and (quantum) gravity in (2+1)-dimensions. Nucl. Phys. B **346**, 440 (1990)
6. Kibble, T.W.B.: Lorentz invariance and the gravitational field. J. Math. Phys. **2**, 212 (1961)
7. Matschull, H.-J.: On the relation between (2+1) Einstein gravity and Chern-Simons theory. Class. Quant. Grav. **16**, 2599 (1999). arXiv:gr-qc/9903040
8. Rovelli, C.: Quantum Gravity, 455p. University Press, Cambridge, UK (2004)
9. Alekseev, A.Y., Malkin, A.Z.: Symplectic structure of the moduli space of flat connection on a Riemann surface. Commun. Math. Phys. **169**, 99 (1995). arXiv:hep-th/9312004
10. Meusburger, C., Schroers, B.J.: Poisson structure and symmetry in the Chern-Simons formulation of (2+1)-dimensional gravity. Class. Quant. Grav. **20**, 2193 (2003). arXiv:gr-qc/0301108
11. Freidel, L., Livine, E.R.: Ponzano-Regge model revisited III: Feynman diagrams and effective field theory. Class. Quant. Grav. **23**, 2021 (2006). arXiv:hep-th/0502106
12. Freidel, L., Livine, E.R.: Effective 3-D quantum gravity and non-commutative quantum field theory. Phys. Rev. Lett. **96**, 221301 (2006). arXiv:hep-th/0512113
13. Meusburger, C., Schroers, B.J.: Quaternionic and Poisson-Lie structures in 3d gravity: the cosmological constant as deformation parameter. J. Math. Phys. **49**, 083510 (2008). arXiv:0708.1507 [gr-qc]

14. Nakahara, M.: Geometry, Topology and Physics, 573p. Taylor & Francis, Boca Raton, USA (2003)
15. Arzano, M., Latini, D., Lotito, M.: Group momentum space and Hopf algebra symmetries of point particles coupled to 2+1 gravity. SIGMA **10**, 079 (2014). arXiv:1403.3038 [gr-qc]
16. Amelino-Camelia, G., Freidel, L., Kowalski-Glikman, J., Smolin, L.: The principle of relative locality. Phys. Rev. D **84**, 084010 (2011). arXiv:1101.0931 [hep-th]
17. Amelino-Camelia, G., Arzano, M., Kowalski-Glikman, J., Rosati, G., Trevisan, G.: Relative-locality distant observers and the phenomenology of momentum-space geometry. Class. Quant. Grav. **29**, 075007 (2012). arXiv:1107.1724 [hep-th]
18. Kowalski-Glikman, J., Trzesniewski, T.: Deformed Carroll particle from 2+1 gravity. Phys. Lett. B **737**, 267 (2014). arXiv:1408.0154 [hep-th]
19. Duval, C., Gibbons, G.W., Horvathy, P.A., Zhang, P.M.: Carroll versus Newton and Galilei: two dual non-Einsteinian concepts of time. Class. Quant. Grav. **31**, 085016 (2014). arXiv:1402.0657 [gr-qc]
20. Bergshoeff, E., Gomis, J., Longhi, G.: Dynamics of Carroll Particles. arXiv:1405.2264 [hep-th]
21. Carroll, L.: Through the Looking Glass and What Alice Found There. MacMillan, London (1871)
22. Amelino-Camelia, G., Lukierski, J., Nowicki, A.: Kappa deformed covariant phase space and quantum gravity uncertainty relations. Phys. Atom. Nucl. **61**, 1811 (1998) [Yad. Fiz. **61**, 1925 (1998)]. arXiv:hep-th/9706031
23. Cianfrani, F., Kowalski-Glikman, J., Pranzetti, D., Rosati, G.: Symmetries of quantum spacetime in three dimensions. Phys. Rev. D **94**(8), 084044 (2016). arXiv:1606.03085 [hep-th]
24. Kowalski-Glikman, J.: κ-Poincaré as a symmetry of flat quantum spacetime. Acta Phys. Polon. Supp. **10**, 321–324 (2017). arXiv:1702.02452 [hep-th]
25. Freidel, L., Geiller, M., Pranzetti, D.: Edge modes of gravity – I: corner potentials and charges. arXiv:2006.12527 [hep-th]
26. Noui, K., Perez, A., Pranzetti, D.: Canonical quantization of non-commutative holonomies in 2+1 loop quantum gravity. JHEP **1110**, 036 (2011). arXiv:1105.0439 [gr-qc]. Noui, K., Perez, A., Pranzetti, D.: Non-commutative holonomies in 2+1 LQG and Kauffman's brackets. J. Phys. Conf. Ser. **360**, 012040 (2012). arXiv:1112.1825 [gr-qc]
27. Pranzetti, D.: Turaev-Viro amplitudes from 2+1 loop quantum gravity. Phys. Rev. D **89**(8), 084058 (2014). arXiv:1402.2384 [gr-qc]
28. Celeghini, E., Giachetti, R., Sorace, E., Tarlini, M.: The three-dimensional Euclidean quantum group E(3)-q and its R matrix. J. Math. Phys. **32**, 1159–1165 (1991). https://doi.org/10.1063/1.529312
29. Amelino-Camelia, G., Smolin, L., Starodubtsev, A.: Quantum symmetry, the cosmological constant and Planck scale phenomenology. Class. Quant. Grav. **21**, 3095–3110 (2004). https://doi.org/10.1088/0264-9381/21/13/002, arXiv:hep-th/0306134 [hep-th]
30. Majid, S., Ruegg, H.: Bicrossproduct structure of kappa Poincare group and noncommutative geometry. Phys. Lett. B **334**, 348 (1994). arXiv:hep-th/9405107

Gravity in 3+1 Dimensions, Particles, and Topological Limit

3

In the previous chapter, we discussed physics in the planar 2+1-dimensional world; now we turn to the physical 3+1-dimensional one. We saw that in 2+1 dimensions, a remarkable thing happens: one can derive the effective particle kinematics by solving out the gravitational degrees of freedom. A natural question arises if also in 3+1 dimensions we have to do with something similar?

There are, clearly, fundamental differences between gravity (coupled to particles) in 2+1 and 3+1 dimensions. First of all, in 3+1 dimensions, gravity is not described by a topological field theory because we know from experience that local gravitational interactions are present. Second, Newton's constant in 3+1 dimensions has the dimension of length over mass $G_N^{(3+1)} = l_P/M_p$, where l_P and M_P are the Planck length and mass, respectively. Therefore, $G_N^{(3+1)}$ itself cannot serve as a deformation parameter, providing a scale for the momentum space geometry. For these reasons, the 3+1-dimensional physics is far more complex than the 2+1-dimensional one, and only partial results are known. Below we will discuss them in detail.

3.1 From 3+1 to 2+1 Dimensions via Dimensional Reduction

Although gravity in 3+1 dimensions differs from gravity in 2+1 dimensions, there are physical circumstances, in which a planar 3+1-dimensional system could be described by the 2+1-dimensional theory. Let us try to construct such a gravitating system, following [1].

The idea is to consider the physical system whose description requires a dimensional reduction from the $3 + 1$-dimensional to the $2 + 1$-dimensional theory. We will do that in quantum theory, using the uncertainty principle as an essential element of the argument. Let us consider a particle in $3 + 1$ dimensions. We assume that the motion of the particle is approximately planar, at least in some classes of

© Springer-Verlag GmbH Germany, part of Springer Nature 2021
M. Arzano and J. Kowalski-Glikman, *Deformations of Spacetime Symmetries*,
Lecture Notes in Physics 986,
https://doi.org/10.1007/978-3-662-63097-6_3

coordinate systems, not accelerating with respect to the natural inertial coordinates at infinity. Let us consider the particle as described by an inertial observer who travels perpendicular to the plane of its motion, which we will call the z direction. From the point of view of the observer, the particle is in an eigenstate of the z-component of the momentum operator, $\hat{P}_z$, with some eigenvalue P_z. Since the particle is in an eigenstate of $\hat{P}_z$, its wavefunction will be uniform in z, with wavelength L. We assume here that L is so large that the standard uncertainty relation can be trusted and therefore

$$L \sim \frac{1}{P_z}. \tag{3.1}$$

At the same time, we assume that the uncertainties in the transverse positions are bounded on a scale r, such that $r \ll 2L$. Then the wavefunction for the particle has support on a narrow cylinder of radius r which extends uniformly in the z direction. Finally, we assume that the state of the gravitational field is semiclassical, so that to a good approximation the system is described by semiclassical Einstein equations

$$G_{ab} = 8\pi G < \hat{T}_{ab} > . \tag{3.2}$$

It should be stressed that we do not have to assume that the semiclassical approximation holds for all states. We assume something much weaker, which is that there are subspaces of states in which it holds and includes the states corresponding to the wavefunction discussed above. This assumption is, in a sense, analogous to the assumption that the spacetime foam effects can be neglected, so that we are close to the ground state of quantum gravity.

Since the wavefunction is uniform in z, we can restrict ourselves to the configurations of the gravitational field possessing a space-like Killing vector field $k^a = (\partial/\partial z)^a$.

Thus, if there are no forces other than the gravitational field, the particle described semiclassically by (3.2) must be described by an equivalent $2 + 1$-dimensional problem in which the gravitational field is dimensionally reduced along the z direction so that the particle, which is the source of the gravitational field, is replaced by a punctures.

The dimensional reduction is governed by a length d, which is the extent in z that the system extends. We cannot take $d < L$ without violating the uncertainty principle. It is then convenient to take $d = L$. Further, since the system consists of the particle, with no intrinsic extent, there is no other scale associated with their extent in the z direction. We can then identify $z = 0$ and $z = L$ to make an equivalent toroidal system, and then dimensionally reduce along z. The relationship between four-dimensional Newton's constant G^4 and three-dimensional Newton's constant $G^3 = G$ is given by

$$G_N^3 \sim \frac{G_N^4}{L} \sim \frac{G_N^4 P_z}{\hbar}. \tag{3.3}$$

Thus, in the analogous three-dimensional system, which is equivalent to the original system as seen from the point of view of the boosted observer, Newton's constant depends on the longitudinal momentum.

Now we note that, if there are no other particles or excited degrees of freedom, the energy of the system can to a good approximation be described by the Hamiltonian H of the two-dimensional dimensionally reduced system. This is described by a boundary integral, which may be taken over any circle that encloses the particle. But we know from discussion in the previous two sections that in $2 + 1$-dimensional gravity, H is bounded from above. This may seem strange, but it is easy to see that it has a natural four-dimensional interpretation.

The bound is given by

$$M \lesssim \frac{1}{4G^3} \sim \frac{L}{4G^4}. \tag{3.4}$$

But this just implies that

$$L \gtrsim 4G^4 M = 2R_{Sch}, \tag{3.5}$$

i.e., this has to be true, otherwise the dynamics of the gravitational field in $3 + 1$ dimensions would have collapsed the system to a black hole. Thus, we see that the total bound from above of the energy in $2 + 1$ dimensions is necessary so that one cannot violate the condition in $3 + 1$ dimensions that a system be larger than its Schwarzschild radius.

Note that we also must have

$$M \gtrsim P_z = \frac{\hbar}{L}. \tag{3.6}$$

Together with (3.5), this implies that L is greater than the Planck length $L \gtrsim l_P$, which is of course necessary if the semiclassical argument we are giving is to hold, and spacetime foam effects are negligible.

Now, we have put no restriction on any components of momentum or position in the transverse directions. So the system still has symmetries in the transverse directions. Furthermore, the argument extends to any number of particles, so long as their relative momenta are coplanar. We can summarize the consequence of the construction presented above as follows.

Let $\mathscr{H}^{QG}$ be the full Hilbert space of quantum theory of gravity, coupled to some appropriate matter fields, with vanishing cosmological constant, $\Lambda = 0$. Let us consider a subspace of states $\mathscr{H}^{weak}$ which are relevant in the low energy limit in which all energies are small in Planck units. We expect that this will have a symmetry algebra which is related to the Poincaré algebra $\mathscr{P}^4$ in 4 dimensions, by some possible small deformations parametrized by G_N^4 and $\hbar$. Let us call this low energy symmetry group $\mathscr{P}_G^4$.

Let us now consider the subspace of $\mathscr{H}^{weak}$ which is described by the system we have just constructed. It contains the particle and is an eigenstate of $\hat{P}_z$ with large P_z and vanishing longitudinal momentum. Let us call this subspace of Hilbert space $\mathscr{H}_{P_z}$.

The conditions that define this subspace break the generators of the (possibly modified) Poincaré algebra that involve the z direction. But they leave unbroken the symmetry in the $2 + 1$-dimensional transverse space. Thus, a subalgebra of $\mathscr{P}_G^4$ acts on this space, which we will call $\mathscr{P}_G^3 \subset \mathscr{P}_G^4$.

We have argued that the physics in $\mathscr{H}_{P_z}$ is a good approximation described by an analogue system of a particle in $2 + 1$-dimensional gravity. However, we know from the results of the last chapter that the symmetry algebra acting there is not the ordinary three-dimensional Poincaré algebra, but its deformation, with the deformation parameter

$$\kappa^{-1} \sim G_N^3 \sim \frac{4G^4 P_z}{\hbar}. \tag{3.7}$$

Therefore, whatever $\mathscr{P}_G^4$ is, it cannot be the standard Poincaré algebra, because it contains a deformed algebra as its subalgebra.

In the following sections, we will use a particularly convenient formulation of $3 + 1$ gravity as a constrained topological field theory to more formally describe a possible $3 + 1$- to $2 + 1$-dimensional reduction outlined above.

3.2   Gravity as a Constrained BF Theory

Although gravity in 3+1 dimensions is not described by a topological field theory, it can be formulated as a topological theory accompanied with a term that breaks the gauge symmetry. Let us show how this comes about. This section follows closely the paper [2] (see also [3]).

As in the previous chapter, we make use of the gauge formulation of gravity, in which the fundamental field is the gauge field $A_\mu = A_\mu^{IJ} T_{IJ}$, and this time with the gauge group $SO(4, 1)$ or $SO(3, 2)$. The generators of the gauge algebra T_{IJ}, $I, J, \ldots = 0, \ldots, 4$ satisfy the algebra

$$[T_{IJ}, T_{KL}] = \eta_{IK} T_{JL} + \eta_{JL} T_{IK} - \eta_{IL} T_{JK} - \eta_{JK} T_{IL}, \tag{3.8}$$

where η_{IJ} is the flat metric $\mathrm{diag}(-1, 1, 1, 1, 1)$ in the case of the algebra $\mathfrak{so}(4, 1)$ and $\mathrm{diag}(-1, 1, 1, 1, -1)$ in the case of the algebra $\mathfrak{so}(3, 2)$. In what follows we will consider only the case of the de Sitter $SO(4, 1)$ gauge group, and the corresponding formulas for the case of $SO(3, 2)$ can be easily derived by changing the sign of the cosmological constant.

The gauge algebra $\mathfrak{so}(4, 1)$ generators T^{IJ} can be decomposed into the translational $P^i \equiv T^{i4}$, $i = 0, \ldots, 3$ and Lorentz T^{ab} ones, satisfying the algebra

$$
\begin{aligned}
[T_{ab}, T_{cd}] &= \eta_{ac} T_{bd} + \cdots, \\
[T_{ab}, T_{c4}] &= \eta_{ac} T_{b4} - \eta_{bc} T_{a4}, \\
[T_{a4}, T_{c4}] &= -\eta_{44} T_{ac},
\end{aligned}
\tag{3.9}
$$

where η_{44} equals 1 in the case of $\mathfrak{so}(4, 1)$ and -1 in the case of $\mathfrak{so}(3, 2)$.

In this chapter, we will make use of the differential forms of formalism in which all the fields are represented by n-forms. This formalism is equivalent to the tensor notation that we used in the previous chapter; however, it makes it possible to write down formulas in a more compact and elegant form. We will write however some of the crucial expressions in components as well.

In accordance with (3.9), the connection one-form (gauge field) $A^{IJ} = A^{IJ}_\mu \, dx^\mu$ can be decomposed into the translational and Lorentz parts

$$A^{a4} = \frac{1}{\ell} e^a , \quad A^{ab} = \omega^{ab} , \tag{3.10}$$

where, as we will see, $e^a \equiv e^a_\mu \, dx^\mu$ is the tetrad one-form (sometimes called the soldering form) and $\omega^{ab} \equiv \omega^{ab}_\mu \, dx^\mu$ is the Lorentz connection (sometimes called the spin connection). Notice that since the gauge field A^{IJ}_μ has the canonical dimension of inverse length, the gauge one-form A^{IJ} is dimensionless; we therefore have to introduce the length scale ℓ to make the tetrad e^a_μ dimensionless.[1] We will see that the scale ℓ is directly related to the cosmological constant.

As usual, the gauge field A^{IJ} is subject to gauge transformations

$$A = A^{IJ} T_{IJ} \to A' = A'^{IJ} T_{IJ} = \left(g^{-1} A g + g^{-1} dg \right)^{IJ} T_{IJ} , \tag{3.11}$$

where it is understood that both the algebra generators $T_{IJ} \equiv (T_{IJ})^a{}_b$ and the group element $g \equiv (g)^a{}_b$ are 5×5 matrices (T_{IJ} is antisymmetric, while g is orthogonal) and the multiplication in the last formula is the matrix multiplication.

Having the gauge field A^{IJ}, we can define the curvature (field strength)

$$F^{IJ} = dA^{IJ} + A^I{}_K \wedge A^{KJ} , \tag{3.12}$$

whose components, using the defining identity $F^{IJ}(A) = \frac{1}{2} F^{IJ}_{\mu\nu} \, dx^\mu \wedge dx^\nu$, can be written as

$$F^{IJ}_{\mu\nu} = \partial_\mu A^{IJ}_\nu - \partial_\nu A^{IJ}_\mu + A^I{}_{\mu K} A^{KJ}_\nu - A^I{}_{\nu K} A^{KJ}_\mu .$$

The curvature splits naturally into the translational and Lorentz components. The former is called torsion and has the form

$$F^{a4} = \frac{1}{\ell} \left(de^a + \omega^a{}_b \wedge e^b \right) \equiv \frac{1}{\ell} D^\omega e^a = \frac{1}{\ell} T^a \tag{3.13}$$

or

$$F^{a4}_{\mu\nu} = \frac{1}{\ell} \left(\partial_\mu e^a_\nu + \omega^a_{\mu b} e^b_\nu - \partial_\nu e^a_\mu - \omega^a_{\nu b} e^b_\mu \right) = \frac{1}{\ell} \left(D^\omega_\mu e^a_\nu - D^\omega_\nu e^a_\mu \right) = \frac{1}{\ell} T^a_{\mu\nu} ,$$

where D^ω is the covariant differential associated with the Lorentz connection ω^{ab}. The latter is called de Sitter curvature and reads

$$F^{ab} = R^{ab} + \frac{1}{\ell^2} e^a \wedge e^b \tag{3.14}$$

---

[1] The tetrad must be dimensionless because the metric $g_{\mu\nu} \equiv \eta_{ab} e^a_\mu e^b_\nu$ is.

or

$$F_{\mu\nu}^{ab} = R_{\mu\nu}^{ab} + \frac{1}{\ell^2}\left(e_\mu^a e_\nu^b - e_\nu^a e_\mu^b\right).$$

In Eq. (3.14), R^{ab} for the standard Lorentz curvature two-form

$$R^{ab}(\omega) = d\omega^{ab} + \omega^a{}_c \wedge \omega_\nu^{cb} \tag{3.15}$$

or

$$R_{\mu\nu}^{ab} = \partial_\mu\omega_\nu^{ab} - \partial_\nu\omega_\mu^{ab} + \omega_{\mu\,c}^a \,\omega_\nu^{cb} - \omega_{\nu\,c}^a \,\omega_\mu^{cb}.$$

This curvature is directly related to the Riemann tensor

$$R_{\rho\sigma\mu\nu} = R_{\mu\nu}^{ab}\, e_{\rho a}\, e_{\sigma b},$$

if the right hand side of this equality is expressed in terms of the metric (which is always possible if torsion vanishes).

The curvature (3.12) transforms homogenously under gauge transformations (3.11), to wit

$$\mathsf{F} = \mathsf{F}^{IJ}\, T_{IJ} \to \mathsf{F}' = \mathsf{F}'^{IJ}\, T_{IJ} = \left(g^{-1}\mathsf{F}g\right)^{IJ} T_{IJ}. \tag{3.16}$$

This gauge invariance is a source of problems when it comes to the construction of the Lagrangian. Indeed we usually demand that the action is gauge and diffeomorphism invariant, but the only geometric, gauge-invariant action that can be constructed from the building blocks that are at our disposal in 3+1 dimensions is just the square of the curvatures

$$L \sim \mathrm{Tr}(\mathsf{F} \wedge \mathsf{F}) \sim \mathsf{F}^{IJ} \wedge \mathsf{F}_{IJ}$$

because, to secure diffeomorphism invariance, the Lagrangian in 3+1 dimensions must be a four-form. But such Lagrangian is a total differential and reduces to the pure boundary term.[2] Therefore, there are no bulk field equations, which means that we have to do with a topological field theory without local interactions. Let us explicitly see how this comes about.

We start decomposing $\mathsf{F}^{IJ} \wedge \mathsf{F}_{IJ}$ into the Lorentz and translational parts

$$\mathsf{F}^{IJ}(A) \wedge \mathsf{F}_{IJ}(A) = R^{ab}(\omega) \wedge R_{ab}(\omega) - \frac{2}{\ell^2}\left(T^a \wedge T_a - R_{ab} \wedge e^a \wedge e^b\right). \tag{3.17}$$

The first term in the expression on the right hand side is the Pontryagin four-form,

$$P_4 = R^{ab} \wedge R_{ab} = 4dC, \tag{3.18}$$

---

[2]One could instead choose the standard Yang–Mills expression $L \sim \mathrm{Tr}(\mathsf{F} \wedge \star\mathsf{F})$, where $\star$ is the Hodge dual, but then the Lagrangian would not be geometrical any longer; it would require the background structure, the metric needed to construct the star $\star$.

where C is the Chern–Simons three-form that we have already encountered in the preceding chapter

$$C = \left(\omega_{ab} \, d\omega^{ab} + \frac{1}{3}\omega_{ab} \wedge \omega^{a}{}_{c} \wedge \omega^{cb} \right) . \tag{3.19}$$

The second term in (3.17) is less known and is called the Nieh–Yan four-form. It can be again expressed as a total differential

$$T^{a} \wedge T_{a} - R_{ab} \wedge e^{i} \wedge e^{j} = 4d \left(e_{a} \wedge D^{\omega} e^{a} \right) . \tag{3.20}$$

In order to construct the Lagrangian for gravity, we will use the following trick. First, we introduce the two-form B^{IJ} valued in the Lie algebra of the gauge group $SO(4, 1)$ that under gauge transformations (3.11), (3.16) transforms as the curvature

$$\mathsf{B} = \mathsf{B}^{IJ} T_{IJ} \rightarrow \mathsf{B}' = \mathsf{B}'^{IJ} T_{IJ} = \left(g^{-1} \mathsf{B} g \right)^{IJ} T_{IJ} . \tag{3.21}$$

Having B and F in our disposal, we construct the most general, gauge and diffeomorphism-invariant Lagrangian

$$L_{top} = \mathsf{B}_{IJ} \wedge \mathsf{F}^{IJ} - \frac{\beta}{2} \mathsf{B}_{IJ} \wedge \mathsf{B}^{IJ} , \tag{3.22}$$

where β is a parameter. Because of the form of the first term, often the name 'BF topological field theory' is used. The theory described by (3.22) is still topological, which can be easily seen by solving B field equations and plugging the solution back to the Lagrangian[3]. Another way to see the topological nature of this theory is to notice that it is invariant not only under the local gauge symmetry, but also under the following symmetry generated by an algebra-valued one-form Φ

$$\mathsf{A}^{IJ} \rightarrow \mathsf{A}'^{IJ} = \mathsf{A}^{IJ} + \Phi^{IJ} ,$$

$$\mathsf{B}^{IJ} \rightarrow \mathsf{B}'^{IJ} = \mathsf{B}^{IJ} + D^{\mathsf{A}} \Phi^{IJ} + \frac{\beta}{2} \left(\Phi^{I}{}_{K} \wedge \Phi^{KJ} + \Phi^{J}{}_{K} \wedge \Phi^{IK} \right) . \tag{3.23}$$

Using this symmetry, one can make $\mathsf{A}'^{IJ} = 0$ and the field equations reduce to trivial $d\mathsf{B}'^{IJ} = 0$.

Now comes the crucial step. We append the Lagrangian (3.22) with an additional term that manifestly breaks the gauge symmetry from de Sitter $SO(4, 1)$ down to

---

[3]If some fields appear in the Lagrangian only algebraically, one can solve their (algebraic) field equations and plug the solution back to the Lagrangian. The field equations obtained from this new Lagrangian are equivalent to the original ones.

its Lorentz subgroup SO(3, 1). To this end, we introduce a constant vector v in the five-dimensional vector space on which the group SO(4, 1) acts[4]

$$v^M = (0, 0, 0, 0, \alpha). \tag{3.24}$$

Obviously, the subgroup of SO(4, 1) that keeps v invariant is the Lorentz group SO(3, 1). With the help of this vector, we can construct an additional term, which is a four-form and is invariant under the reduced SO(3, 1) local gauge symmetry

$$16\pi \, L_{grav} = B_{IJ} \wedge F^{IJ} - \frac{\beta}{2} B_{IJ} \wedge B^{IJ} - \frac{1}{4} \varepsilon_{IJKLM} B^{IJ} \wedge B^{KL} v^M , \tag{3.25}$$

which, taking (3.24) into account, can be rewritten as

$$16\pi \, L_{grav} = B_{IJ} \wedge F^{IJ} - \frac{\beta}{2} B_{IJ} \wedge B^{IJ} - \frac{\alpha}{4} \varepsilon_{IJKL4} B^{IJ} \wedge B^{KL} . \tag{3.26}$$

As we will see in a moment, the Lagrangian (3.25) has the remarkable and unexpected property that it is equivalent to the Einstein–Hilbert–Cartan one.

The Lagrangian (3.25) describes a topological field theory *constrained* by the presence of the gauge symmetry breaking term. Therefore, the theory is called sometimes the *constrained topological field theory*. The last term in (3.24) can be regarded as a small perturbation of the topological field theory. For these reasons, it is hoped that such theory could be an alternative starting point for perturbative quantum gravity, where the diffeomorphism symmetry is manifestly preserved at all steps of perturbation theory (see [2,4] for a more detailed discussion of this point.)

Let us now turn to showing that the Lagrangian (3.25) is indeed a Lagrangian of gravity. To this end, we first solve the field equations of B and plug the solution back to the Lagrangian. We have

$$B^{a4} = \frac{1}{\beta} F^{a4} = \frac{1}{\beta \ell} T^i , \quad B^{ab} = \frac{1}{\alpha^2 + \beta^2} \left(\beta F^{ab} - \frac{\alpha}{2} \varepsilon^{abcd} F_{cd} \right) . \tag{3.27}$$

Plugging this back to the Lagrangian (3.25), we find

$$L_{grav} = \frac{1}{16\pi} \left(\frac{1}{4} M^{abcd} F_{ab} \wedge F_{cd} - \frac{1}{\beta \ell^2} T^i \wedge T_i \right) \tag{3.28}$$

with

$$M^{ab}{}_{cd} = \frac{\alpha}{(\alpha^2 + \beta^2)} \left(\frac{\beta}{\alpha} \delta^{ab}_{cd} - \varepsilon^{ab}{}_{cd} \right) . \tag{3.29}$$

---

[4]In general, v^M can be an arbitrary space-like vector in the flat five-dimensional Minkowski metric of signature $(-, +, +, +, +)$ of norm α^2. Here we have considered a slightly less complicated formulation given by (3.24).

If we now rewrite the parameters α, β, and the scale ℓ as the following functions of the Newton constant G_N, cosmological constant Λ, and the Barbero–Immirzi parameter γ

$$\alpha = \frac{G\Lambda}{3\,(1+\gamma^2)}, \quad \beta = \frac{\gamma G\Lambda}{3\,(1+\gamma^2)}, \quad \gamma = \frac{\beta}{\alpha}, \quad \Lambda = \frac{3}{\ell^2}, \tag{3.30}$$

we obtain the following Lagrangian

$$32\pi G\, L_{grav} = R^{ab} \wedge e^c \wedge e^d\, \varepsilon_{abcd} + \frac{1}{2\ell^2} e^a \wedge e^b \wedge e^c \wedge e^d\, \varepsilon_{abcd} + \frac{2}{\gamma} R^{ab} \wedge e_a \wedge e_b$$

$$+\frac{\ell^2}{2} R^{ab} \wedge R^{cd}\, \varepsilon_{abcd} - \ell^2\gamma\, R^{ab} \wedge R_{ab} + \frac{\gamma^2+1}{\gamma} 2\,(T^a \wedge T_a - R^{ab} \wedge e_a \wedge e_b) \tag{3.31}$$

or, with spacetime indices explicitly written down,

$$64\pi G\, L_{grav} = \varepsilon^{abcd} \left(R_{\mu\nu\,ab} e_\rho c e_\sigma d - \frac{\Lambda}{3} e_\mu a e_\nu b e_\rho c e_\sigma d \varepsilon^{\mu\nu\rho\sigma} + \frac{2}{\gamma} R_{\mu\nu\,ab}\, e_\nu^a e_\rho^b \right) \varepsilon^{\mu\nu\rho\sigma}$$

$$+ \frac{\gamma^2+1}{\gamma} NY_4 + \frac{3\gamma}{2\Lambda} P_4 - \frac{3}{4\Lambda} E_4. \tag{3.32}$$

The first term in (3.31) is the Einstein–Hilbert action in the first-order form and the second is the cosmological constant term. The third term is called the Holst term [5]. This term is not topological, but it does not influence the theory if torsion vanishes, or even in some cases when torsion is non-trivial, e.g., in supergravity [6].

The field equations can be conveniently derived from the Lagrangian (3.31) and turn out to be Einstein equations with cosmological constant, as expected. Let us however take a step back and try to derive them from the variation of the Lagrangian (3.26). We find the following field equations

$$(D^A B)^{IJ} = 0, \tag{3.33}$$

$$F^{IJ} - \beta B^{IJ} - \frac{\alpha}{2} \varepsilon^{IJKL4} B_{KL} = 0, \tag{3.34}$$

where D^A is the covariant derivative defined by the connection A^{IJ}, defined by

$$(D^A B)^{IJ} = dB^{IJ} + A^I{}_K \wedge B^{KJ} + A^J{}_K \wedge B^{IK}. \tag{3.35}$$

Let us solve Eqs. (3.33) and (3.34). This task simplifies a lot if we consider first an integrability condition for (3.34) obtained acting with the covariant differential D^A on it. To calculate the resulting expression, it is helpful to reintroduce the vector v^M and rewrite (3.34) as

$$F^{IJ} - \beta B^{IJ} - \frac{\alpha}{2} \varepsilon^{IJKLM} B_{KL} v_M = 0.$$

Acting with D^A on this equation, and using the fact that $D^A F = 0$ (Bianchi identity), (3.33), and the fact that the covariant differential of the ε symbol vanishes, we get

$$0 = \varepsilon^{IJKLM} B_{KL} \wedge D^A v_M = -\varepsilon^{IJKLm} B_{KL} \wedge e_m , \qquad (3.36)$$

because

$$D^A v^M = dv^M + A^M{}_N v^N = A^M{}_4 .$$

If $I, J \neq 4$ and $L = 4$, we find the condition

$$\varepsilon^{abcd} T_c \wedge e_d = 0 . \qquad (3.37)$$

It can be shown that, under the condition that the tetrad e^i_μ is invertible, the only solution of this equation is that torsion vanishes

$$T_a = 0 . \qquad (3.38)$$

Let us now assume that $I = 4$ in which case we have

$$\varepsilon^{abcd} B_{bc} \wedge e_d = 0 . \qquad (3.39)$$

From (3.27), we know that the expression for B has the symbolic form

$$B_{bc} \sim F_{bc} + \varepsilon_{bc}{}^{ef} F_{ef} .$$

Let us first show that the second term in the above expression does not contribute to (3.39). Indeed

$$\varepsilon^{abcd} B_{bc} \wedge e_d \sim F_{ef} \wedge e^e \sim R_{ef} \wedge e^e \equiv 0 ,$$

where in the last step, we used the identity $R_{ab} \wedge e^a \equiv (D^\omega T)_b$. Returning to (3.39), we see that we are left with

$$\varepsilon^{abcd} F_{bc} \wedge e_d = \varepsilon^{abcd} \left(R_{bc} \wedge e_d + \frac{\Lambda}{3} e_b \wedge e_c \wedge e_d \right) = 0 , \qquad (3.40)$$

which are Einstein equations in the presence of the cosmological constant Λ.

This concludes our presentation of pure gravity. Let us now turn to the particle coupling.

3.3 Particles Coupled to Gravity

In the preceding section, we presented the construction of the theory of gravity in 3+1 spacetime dimensions as a constrained BF theory, with the BF gauge group SO(4, 1) (or SO(3, 2)) broken down to the Lorentz subgroup SO(3, 1). Now it is time to consider the matter sources of gravity. As we have seen in the previous chapter, in 2+1 spacetime dimensions, matter can be described by introducing the simplest possible term breaking the gauge symmetry of the theory in a localized way, along the particle worldline. This can be interpreted also as the insertion of a Wilson line in the spacetime [4]. The gauge degrees of freedom are then promoted to dynamical degrees of freedom and reproduce the dynamics of a relativistic particle coupled to gravity. In the four-dimensional case, we can work along similar lines introducing matter (relativistic particles) as charged (under SO(4, 1)) topological gravitational defects. This treatment provides a new perspective in which matter and gravity are geometrically unified. In this picture, gravity is the geometry of the spacetime manifold with defects describing matter.

The construction of the particle Lagrangian is exactly the same as it was in the case of 2+1 gravity, so we start with recalling the relevant expressions (see (2.28) and (2.29)) (with suppressed SO(4, 1) indices)

$$L_{part} = \left(\langle h^{-1}(x)\dot{h}(x)\,\mathscr{C} \rangle + \langle A_\tau(x)\,h(x)\,\mathscr{C}h(x)^{-1} \rangle \right) \delta(x - z(\tau)), \qquad (3.41)$$

where the inner product $\langle * \rangle$ is understood to be the matrix trace,

$$\langle T_{IJ}T_{KL} \rangle = \text{Tr}\,(T_{IJ}T_{KL}) = \eta_{IK}\eta_{JL} - \eta_{JK}\eta_{IL}, \qquad (3.42)$$

while the rest mass and spin are described by the Cartan subalgebra of the SO(4, 1) algebra element $\mathscr{C}$ of the form

$$\mathscr{C} = m\ell\,T^{04} + s\,T^{23}, \qquad (3.43)$$

$z(\tau)$ is the particle worldline, and $A_\tau \equiv A_\mu \dot{z}^\mu$.

Let us first show that the equations of motion of the particle are the correct Mathisson–Papapetrou equations [8] describing the particle with mass and spin moving on the gravitational background, generalized by the possible presence of the non-trivial torsion. To this end, we first take $h(x)$ to be an element of the Lorentz subgroup SO(3, 1) of SO(4, 1). This choice is justified by the fact that, as discussed in the preceding section, it is only the Lorentz group that is a gauge group of gravity. We will return to the case where $h \in$ SO(4, 1) below.

In the first step, let us introduce a convenient notation. Let

$$\mathscr{J} = h\,\mathscr{C}\,h^{-1} \equiv \ell\,p_a\,T^{a4} + \frac{1}{2}\,s_{ab}\,T^{ab}. \qquad (3.44)$$

It follows from this definition that the momentum p_a satisfies the standard, relativistic mass-shell condition

$$p_0^2 - \mathbf{p}^2 = m^2. \qquad (3.45)$$

It can be also checked that the momentum is orthogonal to the spin

$$p_a \, s^{ab} = 0 . \qquad (3.46)$$

Indeed, this equation is Lorentz covariant and therefore must hold, in particular, in the particle's rest frame. But in this frame (cf. (3.43)), the only nonvanishing component of the momentum p_a is $p_0 = m$ while the only nonvanishing component of the spin s^{ab} is $s^{23} = s$. This proves (3.46).

The variation of the Lagrangian (3.41) with respect to h gives (neglecting total derivatives)

$$\delta L = - \langle \delta h \, h^{-1} \left(\dot{\mathscr{J}} + [A_\tau, \, \mathscr{J}] \right) \rangle . \qquad (3.47)$$

Since, by assumption, h belongs to the Lorentz subgroup $SO(3, 1)$, only the Lorentz component of the expression in parentheses in (3.47) is picked up, and therefore the resulting equations of motion are

$$\dot{\mathscr{J}}^{ab} + A^a_{\tau K} \, \mathscr{J}^{Kb} + A^b_{\tau K} \, \mathscr{J}^{aK} = 0 \qquad (3.48)$$

or

$$D^\omega_\tau \, \mathscr{J}^{ab} - e^a_\tau \, p^b + e^b_\tau \, p^a = 0 , \qquad (3.49)$$

where

$$D^\omega_\tau \, \mathscr{J}^{ab} = \dot{\mathscr{J}}^{ab} + \omega^a_{\tau c} \, \mathscr{J}^{cb} + \omega^b_{\tau c} \, \mathscr{J}^{ac} . \qquad (3.50)$$

Equation (3.48) relates the time evolution of spin with the momentum of the particle and is called the spin precession equation.

Let us now consider the equation that results from the variation of the particle Lagrangian (3.41) with respect to the particle trajectory z^μ. There are two kinds of contributions to this variation. The first comes from varying the argument of the group element $h(z(\tau))$ but this variation produces just a specific kind of the variation $\delta h = \partial h / \partial z \, \delta z$ multiplied by (3.48), and therefore it leads to nothing new. The new equation is produced by variation of $A_\mu(z(\tau)) \, \dot{z}^\mu$ which gives

$$\frac{\delta L}{\delta z^\mu} = \left\langle \frac{d}{d\tau} \left(\mathscr{J} A_\mu \right) - \mathscr{J} \, \partial_\mu A_\nu \dot{z}^\nu \right\rangle = \left\langle D^A_\tau \, \mathscr{J} A_\mu - \mathscr{J} F_{\mu\nu}(A) \dot{z}^\nu \right\rangle = 0 . \qquad (3.51)$$

Using (3.48), we see that (3.51) reduces to

$$2 \left(D^A_\tau \, \mathscr{J} \right)_{a4} A^{a4}_\mu - \mathscr{J}_{IJ} F^{IJ}_{\mu\nu}(A) \dot{z}^\nu = 0, \qquad (3.52)$$

which, when rewritten in components, gives

$$\left(D^\omega_\tau \, p_a \right) e_\mu{}^a = \frac{1}{2} \, s_{ab} \, R_{\mu\nu}{}^{ab} \, \dot{z}^\nu + p_a T_{\mu\nu}{}^a \, \dot{z}^\nu . \qquad (3.53)$$

This is the Mathisson–Papapetrou equation describing the motion of the massive, spinning particle in the presence of torsion. When torsion vanishes (which is, strictly speaking, not consistent, because, as we will see in the next section, the spin of the particle is a source of torsion), we obtain the standard Mathisson–Papapetrou equation; if spin also vanishes, we get the standard geodesic equation.

Equation (3.53) can be rewritten in a more standard form introducing the affine connection $\hat{\Gamma}_{\mu\nu}{}^\rho$, related to the spin connection ω_μ^{ab} by

$$\partial_\mu e_\nu{}^a + \omega_\mu{}^a{}_b e_\nu^b = \hat{\Gamma}_{\nu\mu}{}^\rho e_\rho{}^a \,. \tag{3.54}$$

In the presence of torsion, the affine connection depends on the Christoffel symbols $\Gamma_{\mu\nu}{}^\rho$ and torsion as follows:

$$\hat{\Gamma}_{\mu\nu\rho} = \Gamma_{\mu\nu\rho} + T_{\rho\{\mu\nu\}} - \frac{1}{2}T_{\mu\nu\rho}, \tag{3.55}$$

where the torsion tensor is given by

$$T_{\mu\nu\rho} = T_{\mu\nu}{}^a e_{\rho a} \,. \tag{3.56}$$

Expressed in terms of this affine connection, the Mathisson–Papapetrou equation reads

$$\nabla_\tau p_\mu = \frac{1}{2} s_{ab} R_{\mu\nu}{}^{ab} \dot{z}^\nu \,, \tag{3.57}$$

with

$$p_\mu \equiv p_a e_\mu{}^a$$

and

$$\nabla_\mu p_\nu \equiv \partial_\mu p_\nu - \Gamma_{\mu\nu}{}^\rho p_\rho$$

being the covariant derivative.

Before closing this section, let us return to our assumption that the group element h belongs to the Lorentz subgroup SO(3, 1) of the gauge group SO(4, 1). What would go wrong if we relax it? In that case, in addition to (3.48) from variation over h, we would get an equation for components $\mathscr{I}^{a4}$, which has the form

$$D_\tau^\omega p_a = \frac{1}{\ell^2} s_{ab} e_\tau{}^b \,. \tag{3.58}$$

If we compare this equation with (3.53), we find that the following condition must be satisfied along the particle's worldline

$$\frac{1}{\ell^2} s_{ab} e_\mu{}^a e_\tau{}^b = \frac{1}{2} s_{ab} R_{\mu\nu}{}^{ab} \dot{z}^\nu + p_a T_{\mu\nu}{}^a \dot{z}^\nu \,. \tag{3.59}$$

This last equation has to be satisfied for an arbitrary particle's spin and momentum, and it follows that

$$R_{\mu\nu}{}^{ab} - \frac{1}{l^2} e_\mu{}^a e_\tau{}^b = 0, \quad T_{\mu\nu}{}^a = 0, \tag{3.60}$$

i.e., only if the background geometry is de Sitter, which is clearly unacceptable. As we will see in the next section, this problem does not arise if we consider the full gravity–particle system, i.e., if we take into account the backreaction of the particle on geometry.

3.4 Equations for the Particle–Gravity System

Let us now turn to consider the gravitational field equations in the case when the gravitational field is coupled to a particle carrying the charge $\mathscr{J}$ (3.44). The Lagrangian of this system is given by the sum of (3.25) and (3.41)

$$L = \mathsf{B}_{IJ} \wedge \mathsf{F}^{IJ} - \frac{\beta}{2} \mathsf{B}_{IJ} \wedge \mathsf{B}^{IJ} - \frac{1}{4} \varepsilon_{IJKLM} \mathsf{B}^{IJ} \wedge \mathsf{B}^{KL} v^M +$$
$$\int d\tau \left(\langle h^{-1}(x)\dot{h}(x)\, \mathscr{C} \rangle + \langle \mathsf{A}_\tau(x)\, h(x)\, \mathscr{C} h(x)^{-1} \rangle \right) \delta^4(x - z(\tau)). \tag{3.61}$$

The equations of motion resulting from this Lagrangian are as follows.

B equations.
After decomposing into Lorentz and translational parts, we have

$$B^{ab} = \frac{1}{\alpha^2 + \beta^2} \left[-\frac{\alpha}{2} \varepsilon^{abcd} F_{cd} + \beta F^{ab} \right] \tag{3.62}$$

and

$$B^{a4} \equiv B^a = \frac{1}{\beta} F^{a4} = \frac{1}{\beta\ell} D^\omega e^a. \tag{3.63}$$

These equations are algebraic and express B in terms of the curvature F.

A equations.

$$\left(D^A B \right)^{IJ} = \frac{1}{2} \mathscr{J}_P^{IJ}(x), \tag{3.64}$$

where we have introduced the three-form

$$\mathscr{J}_P^{IJ}(x) = \int \varepsilon_{\mu\nu\rho\sigma} \mathscr{J}^{IJ}(\tau)\dot{z}^\sigma \delta^4(x - z(\tau)) dx^\mu \wedge dx^\nu \wedge dx^\rho. \tag{3.65}$$

Here $\varepsilon_{\mu\nu\rho\sigma}$ is the Levi-Civita tensor $\varepsilon_{0123} = 1$. The form $\mathscr{J}_P^{IJ}(x)$ satisfies

$$\int \mathscr{J}_P^{IJ}(x) \wedge a(x) = \int_P d\tau \, \mathscr{J}^{IJ}(\tau) \dot{z}^{\mu}(\tau) a_{\mu} \qquad (3.66)$$

for any one-form $a = a_{\mu} dx^{\mu}$.

h equations.

Contrary to what we did in the previous section, the equation below is obtained by varying the action with respect to $h \in \mathsf{SO}(4, 1)$.

$$\alpha \, \varepsilon^{abcd} \, \mathsf{B}_{ab} \wedge \mathsf{B}_{c4} = \left(\int d\tau \, (D_{\tau}^{\mathsf{A}} \, \mathscr{J})^d \, \delta^4(x - z(\tau)) \right) d^4 x. \qquad (3.67)$$

This completes the full set of equations.

Notice that in the case of a particle in an external, fixed, gravitational field, the inclusion of the h equation of motion for $h \in \mathsf{SO}(4, 1)$ (3.58) leads to constraints imposed on the components of gravitational field strengths. As we will see in a moment, this problem is absent if the gravitational field is dynamical, in the case $\alpha \neq 0$.

Let us indeed consider Eq. (3.64). Applying the covariant derivative to both sides, we get

$$\left(D^{\mathsf{A}} \, D^{\mathsf{A}} \, \mathsf{B} \right)^{IJ} = \frac{1}{2} (D^{\mathsf{A}} \, \mathscr{J}_P)^{IJ}(x) = -\frac{1}{2} \left(\int D_{\tau} \, \mathscr{J}(\tau) \, \delta^4(x - z(\tau)) \, d\tau \right)^{IJ}.$$
$$(3.68)$$

Therefore, the component $a4$ of (3.68) is just $-1/2$ times the RHS of Eq. (3.67). On the other hand, from the Bianchi identity, we have

$$\left(D^{\mathsf{A}} \, D^{\mathsf{A}} \, \mathsf{B} \right)^{IJ} = \mathsf{F}^I{}_K \wedge \mathsf{B}^{KJ} + \mathsf{F}^J{}_K \wedge \mathsf{B}^{IK}.$$

To compare this with Eq. (3.67), we just need to extract the translational component

$$\mathsf{F}^d{}_c \wedge \mathsf{B}^{c4} + \mathsf{F}^4{}_c \wedge \mathsf{B}^{dc} = \left(\mathsf{B}^{cd} - \frac{1}{\beta} \mathsf{F}^{cd} \right) \wedge \frac{1}{\ell} D^{\omega} e_c$$

$$= -\frac{\alpha}{\beta} \frac{1}{\alpha^2 + \beta^2} \left(\alpha \, \mathsf{F}^{cd} + \frac{\beta}{2} \varepsilon^{cdab} \mathsf{F}_{ab} \right) \wedge \frac{1}{\ell} D^{\omega} e_c \qquad (3.69)$$

$$= -\frac{\alpha}{2} \varepsilon^{abcd} (\mathsf{B}_{ab} \wedge \mathsf{B}_{c4}).$$

This is just $-1/2$ times the LHS of Eq. (3.67). Thus, we conclude that the h Eq. (3.67) is just a part of the integrability conditions of (3.64). Thus, in what follows, we can disregard Eq. (3.67) whatsoever. This surprising at the first sight conclusion can be understood if one recalls that the fact that the matter field equation that follows from the integrability condition for Einstein equations is a well-known property of Einstein's theory of gravity.

We see therefore that in the case, in which gravity is fully dynamical, it is consistent to take the gauge degrees of freedom that become dynamical at the particle world-line, described by $h \in SO(4, 1)$. Diffeomorphisms and those group elements h that belong to $SO(3, 1)$ leave the bulk action invariant, and therefore they are dynamical degrees of freedom only along the worldline location. Those h that belong to $SO(4, 1)/SO(3, 1)$ do not leave the bulk action invariant, and therefore they are dynamical degrees of freedom even in the absence of the particle. Since their equation of motion is a subset of the Einstein equation, this suggests that they are determined on-shell by the gravitational and particle degree of freedom.

Let us now consider the equation for the gravitational field produced by a point particle in full generality. Our starting point will be Eqs. (3.62), (3.63), and (3.64). Let us consider first the Lorentz components of Eq. (3.64) and expand its RHS.

$$
\begin{aligned}
\left(D^A B\right)^{ab} &= D^\omega B^{ab} + A^{a4} \wedge B_4{}^b + A^{b4} \wedge B^a{}_4 \\
&= D^\omega B^{ab} + \frac{1}{\beta \ell^2} \left(T^a \wedge e^b - e^a \wedge T^b\right) .
\end{aligned} \tag{3.70}
$$

Now, from definition of the curvature F, (3.14)

$$
D^\omega F^{ab} = D^\omega R^{ab} - \frac{1}{\ell^2} \left(T^a \wedge e^b - T^b \wedge e^a\right) = -\frac{1}{\ell^2} \left(T^a \wedge e^b - T^b \wedge e^a\right). \tag{3.71}
$$

Thus

$$
\left(D^A B\right)^{ab} = D^\omega (B^{ab} - \frac{1}{\beta} F^{ab}) = -\frac{\alpha}{\beta} \frac{1}{\alpha^2 + \beta^2} D^\omega \left(\alpha F^{ab} + \frac{\beta}{2} \varepsilon^{abcd} F_{cd}\right) = \frac{1}{2} \mathscr{J}_P^{ab}
$$

and Eq. (3.64) can be written as

$$
T^a \wedge e^b - T^b \wedge e^a = \frac{\beta \ell^2}{2\alpha} \left(\alpha \delta^{ab}_{cd} - \frac{\beta}{2} \varepsilon^{ab}{}_{cd}\right) \mathscr{J}_P^{cd} . \tag{3.72}
$$

This is an algebraic equation which fully determines the torsion in terms of the spin of the particle. When written in terms of the gravitational constant, it reads

$$
T^a \wedge e^b - T^b \wedge e^a = \frac{G\gamma}{2(1 + \gamma^2)} \left(\delta^{ab}_{cd} - \frac{\gamma}{2} \varepsilon^{ab}{}_{cd}\right) s_P^{cd}, \tag{3.73}
$$

and we see that the Barbero–Immirzi parameter γ affects the coupling between torsion and spin; in the case of usual metric gravity $\gamma = 0$, the torsion is zero.

If we now consider the translation part of (3.64), we get

$$
\begin{aligned}
\left(D^A B\right)^{a4} &= D^\omega B^{a4} + A^4{}_b \wedge B^{ab} \\
&= \frac{1}{\ell}\left(\frac{1}{\beta} D^\omega T^a - B^{ab} \wedge e_b\right) \\
&= \frac{1}{\ell}\left(\frac{1}{\beta} F^{ab} - B^{ab}\right) \wedge e_b .
\end{aligned}
\tag{3.74}
$$

Thus, the translational part of the field equations gives us the Einstein equations

$$
\frac{\alpha}{(\alpha^2 + \beta^2)}\left(\frac{\alpha}{\beta} R^{ab} \wedge e_b + G^a\right) = \frac{\ell}{2}\mathcal{J}_P^{a4},
\tag{3.75}
$$

where

$$
G^a \equiv \frac{1}{2}\varepsilon^{abcd} F_{cd} \wedge e_b = \frac{1}{2}\varepsilon^{abcd}\left(R_{cd} \wedge e_b - \frac{\Lambda}{3} e_b \wedge e_c \wedge e_d\right)
$$

is the Einstein tensor with cosmological constant. The first term on the RHS is the derivative of the torsion $D^{\omega 2} e^a = D^\omega T^a = R^{ab} \wedge e_b$. Equation (3.75) written in terms of the gravity coupling constant (cf. (3.30)) reads

$$
\frac{1}{\gamma} R^{ab} \wedge e_b + G^a = \frac{G}{2} p_P^a.
\tag{3.76}
$$

Equations (3.73) and (3.76) characterize the gravitational field produced by point particle with momentum p^a and spin s^{ab} and have the standard form.

3.5 The Topological Vacuum

As we saw in the last two sections, gravity in 3+1 dimensions, in spite of possessing local degrees of freedom, is in some sense similar to gravity in 2+1 dimensions discussed in the previous chapter; while the latter is described by a topological field theory, the former is given by topological field theory with constraints. It is tempting therefore to treat the exact solutions of the BF topological field theory in 3+1 dimensions as vacuum states of gravity and regard the constraint term proportional to the parameter α in the Lagrangian (3.25) as a small perturbation around this vacuum. It follows from (3.30) that this case corresponds to the theory with large Barbero–Immirzi parameter γ, so that $\alpha/\beta = \gamma^{-1} \to 0$.

There is a great advantage, in principle, of such perturbations around the topological vacuum, in powers of the parameter α as compared to the standard perturbative expansion around the gravitational vacuum, usually assumed to be the Minkowski space. As it is well known (see, e.g., [9]), this standard perturbation theory manifestly breaks the diffeomorphism invariance of general relativity. Contrary to that, the perturbative expansion around the topological vacuum manifestly preserves diffeomorphism invariance.

Let us start with writing down the field equations obtained in the last section

$$(D^A B)^{IJ} = \mathscr{J}_P^{IJ}(x) \tag{3.77}$$

and

$$F^{IJ} = \alpha\, \varepsilon^{IJKLM} B_{JK}\, v_M + \beta B^{IJ}, \tag{3.78}$$

where in the last equation, we reintroduced the gauge breaking vector $v^M = (0, 0, 0, 0, 1)$. We want to solve these equations perturbatively, in powers of α. To this end, we expand

$$A = A^{(0)} + \alpha\, A^{(1)} + \cdots, \quad B = B^{(0)} + \alpha\, B^{(1)} + \cdots. \tag{3.79}$$

At zeroth order in α, we get the field equations describing the particle coupled to the BF topological field theory

$$(D^{A^{(0)}} B^{(0)})^{IJ} = \mathscr{J}_P^{IJ}(x) \tag{3.80}$$

and

$$F^{IJ}(A^{(0)}) = \beta B^{(0)IJ}. \tag{3.81}$$

As it turns out, the theories obtained by putting α strictly equal to zero and the one with a nonzero, but small, α are fundamentally different. In the first case, we have just Eqs. (3.80) and (3.81); in the second, we have to check if there are no integrability conditions for the higher order equations which involve only the topological background fields $A^{(0)}$ and $B^{(0)}$. As we will see in a moment, this is exactly what is happening. Indeed let us write down the first-order equations

$$\left(D^{A^{(0)}} A^{(1)}\right)^{IJ} = \varepsilon^{IJKLM} B^{(0)}{}_{JK}\, v_M + \beta B^{(1)IJ}, \tag{3.82}$$

$$D^{A^{(0)}} B^{(1)IJ} + A^{(1)I}{}_K \wedge B^{(0)KJ} + A^{(1)J}{}_K \wedge B^{(0)IK} = 0. \tag{3.83}$$

If we now apply the covariant differential $D^{A^{(0)}}$ to Eq. (3.82), use (3.78), and compare it with (3.83), we find the following condition:

$$\varepsilon^{IJKLM} D^{A^{(0)}} (B^{(0)}{}_{JK}\, v_M) = 0. \tag{3.84}$$

One can check that there are no integrability conditions resulting from the higher order equations (of order $\alpha^2, \alpha^3, \ldots$) involving only the zeroth-order fields. Therefore, in the case of α small and nonvanishing, contrary to the case of the BF theory coupled to the particle, the set of field equations contains (3.80), (3.81), and (3.84).

If we try to solve these equations, a seemingly grave problem immediately arises. Substituting (3.81) into (3.80), we find that the left hand side has the form $1/\beta\, D^{A^{(0)}} F(A^{(0)})$ which identically vanishes by virtue of the Bianchi identity. It seems therefore that it is not possible to couple any non-trivial source to the theory. To resolve this puzzle, let us notice that the Bianchi identity holds only for curvatures of smooth connections, and in particular, the derivation of this identity explicitly makes use of the fact that $dd\mathsf{A} = 0$. Therefore, we take

$$\mathsf{A}^{IJ} = \mathsf{A}^{IJ}_{reg} + \beta\,\mathscr{G}^{IJ}\,\mathfrak{a}_D\,, \tag{3.85}$$

where A^{IJ}_{reg} is a smooth connection, while, assuming that the particle is at the origin of the cartesian coordinates $\mathbf{x} = (x^1, x^2, x^3)$,

$$dd\mathfrak{a}_D = \delta^3(\mathbf{x})\,dx^1 \wedge dx^2 \wedge dx^3\,. \tag{3.86}$$

One easily checks that

$$\mathfrak{a}_D = \frac{1}{4\pi}\,(1 - \cos\theta)\,d\phi \tag{3.87}$$

satisfies (3.86), because integrating the left hand side of this equation over the unit ball B

$$\int_B dd\mathfrak{a}_D = \int_{S^2} d\mathfrak{a}_D = 1\,.$$

Therefore, the particle coupled to the BF topological field theory behaves like a Dirac magnetic monopole (see, for example, [10]), it has the Dirac string attached, being the line singularity at $\theta = 0$ of the connection (3.87). In fact, it is the Taub-NUT space (gravitational monopole) that is the solution of the field equation in this case. More details and discussion can be found in [7]. It is tempting to speculate that the Taub-NUT solution serves as a topological vacuum for gravity coupled to a particle and that the local degrees of freedom of gravity described by the α gauge symmetry breaking term in the gravity action arise somehow dynamically, as a result of some fundamental principle. Some investigations along these lines in the context of entropic gravity were presented in [11].

3.6   Effective Deformed Particle

Let us investigate the particle coupled to topological gravity from a different perspective [12]. We know that in the topological limit $\alpha = 0$, the constrained BF theory (3.25) becomes topological

$$L_{grav} = B_{IJ} \wedge F^{IJ} - \frac{\beta}{2}\,B_{IJ} \wedge B^{IJ}\,. \tag{3.88}$$

Solving for the field B, we find

$$B^{IJ} = \frac{1}{\beta} F^{IJ} \tag{3.89}$$

and substituting it back to the Lagrangian, we get

$$L_{topgrav} = \frac{1}{2\beta} F_{IJ} \wedge F^{IJ} , \tag{3.90}$$

where we reabsorbed the parameter β into redefinition of the gravitational constant. To proceed, let us rewrite the Lagrangian (3.90) in tensorial form

$$L_{topgrav} = \frac{1}{2\beta} \varepsilon^{\mu\nu\rho\sigma} \langle F_{\mu\nu} F_{\rho\sigma} \rangle \tag{3.91}$$

with the summation over the gauge indices replaced by the inner product $\langle \star \rangle$, with

$$F_{\mu\nu}^{IJ} = \partial_\mu A_\nu^{IJ} + A_\mu^I{}_K A_\nu^{KJ} - \mu \leftrightarrow \nu. \tag{3.92}$$

One checks by direct computation that the topological Lagrangian $L_{topgrav}$ is a total derivative of the Chern–Simons one

$$L_{topgrav} = \frac{1}{2\beta} \varepsilon^{\mu\nu\rho\sigma} \partial_\sigma \left\langle A_\mu \partial_\nu A_\rho + \frac{2}{3} A_\mu A_\nu A_\rho \right\rangle . \tag{3.93}$$

Therefore, for any spacetime bounded region, the action of topological gravity contains only the three-dimensional boundary term and becomes the action of 2+1 gravity there.

Let us now consider the topological gravity coupled to a single particle. The total Lagrangian will be in this case the sum of (3.91) and (3.41)

$$L_{part} = \left(\langle h^{-1}(x)\dot{h}(x)\, \mathscr{C} \rangle + \langle A_\tau(x)\, h(x)\, \mathscr{C} h(x)^{-1} \rangle \right) \delta(x - z(\tau)). \tag{3.94}$$

Let us assume that the particle world line belongs to some three-dimensional surface $\mathscr{S} = \Sigma \times \mathbb{R}$, being a product of the two-dimensional space-like surface Σ and the time axis $\mathbb{R}$. Taking then the theory defined on the four-dimensional volume $\mathscr{V}$ with boundary $\mathscr{S}$ with the action

$$S = \int_{\mathscr{V}} d^4x \, L_{topgrav} + L_{part} \tag{3.95}$$

we see that this theory is effectively described by the Chern–Simons action (2.49) that we are already well acquainted with. To what follows, it is convenient to write the Lagrangian using differential forms with $A_\Sigma = A_i dx^i$, to wit

$$L = \frac{k}{4\pi} \int_\Sigma \langle \dot{A}_\Sigma \wedge A_\Sigma \rangle - \langle \mathscr{C} h^{-1}\dot{h} \rangle + \int \left\langle A_0, \frac{k}{2\pi} F_\Sigma - h\mathscr{C} h^{-1} \delta^2(\mathbf{x})\, dx^1 \wedge dx^2 \right\rangle , \tag{3.96}$$

with spatial curvature $F_\Sigma = dA_\Sigma + [A_\Sigma, A_\Sigma]$. The gauge group is now $SO(4, 1)$ with the inner product being

$$\langle T_{IJ} T_{KL} \rangle = \text{Tr}\,(T_{IJ} T_{KL}) = \eta_{IK}\eta_{JL} - \eta_{JK}\eta_{IL}\,, \tag{3.97}$$

and the Cartan subalgebra element defining the particle at rest has now the form

$$\mathscr{C} = m\ell\,T^{04} + s\,T^{23}\,. \tag{3.98}$$

We can follow the steps described in Sect. 2.3 (see also in-depth discussion in [13]), where we solved the constraint and plugged the solution back to the action to obtain the effective deformed action of the particle. The only difference between the present case and the one considered there is that now we have to do with a different gauge group, which is going to be of importance in a moment. We solve the constraint in exactly the same way as we did before in Sect. 2.3. We decompose the manifold Σ into the plaquette $\mathscr{D}$ and the asymptotic region $\mathscr{H}$, as in Fig. 2.1. In these regions, the solution of the constraint

$$\frac{k}{2\pi}F_\Sigma - h\mathscr{C}h^{-1}\delta^2(\mathbf{x})\,dx^1 \wedge dx^2 \tag{3.99}$$

has the form

$$A_\Sigma^{\mathscr{H}} = \gamma^{-1}d\gamma\,, \tag{3.100}$$

and

$$A_\Sigma^{\mathscr{D}} = \frac{1}{k}\bar\gamma^{-1}\mathscr{C}\bar\gamma d\phi + \bar\gamma^{-1}d\bar\gamma\,, \quad \bar\gamma(0) = h^{-1}\,. \tag{3.101}$$

Plugging these solutions to the Lagrangian (3.96) and using the identity $dd\phi = 2\pi\,\delta^2(\mathbf{x})\,dx^1 \wedge dx^2$, we obtain

$$
\begin{aligned}
L = {} & \frac{k}{4\pi}\int_\Gamma \left\langle \dot\gamma\gamma^{-1}d\gamma\gamma^{-1} - \dot{\bar\gamma}\bar\gamma^{-1}d\bar\gamma\bar\gamma^{-1} + \frac{2}{k}\mathscr{C}d\phi\,\dot{\bar\gamma}\bar\gamma^{-1} \right\rangle \\
& + \frac{k}{4\pi}\int_\mathscr{H} \left\langle \dot\gamma\gamma^{-1}d\gamma\gamma^{-1} \wedge d\gamma\gamma^{-1} \right\rangle \\
& + \frac{k}{4\pi}\int_\mathscr{D} \left\langle \dot{\bar\gamma}\bar\gamma^{-1}d\bar\gamma\bar\gamma^{-1} \wedge d\bar\gamma\bar\gamma^{-1} \right\rangle,
\end{aligned} \tag{3.102}
$$

where Γ is the common boundary of $\mathscr{D}$ and $\mathscr{H}$. In the second and third line of the above expression, we have the bulk contributions called Wess-Zumino–Witten term or WZW term for short. These terms do not influence the deformed particle dynamics and we disregard them in what follows.

Now we must impose the continuity condition along the boundary Γ

$$\left. A^{\mathscr{D}} \right|_\Gamma = \left. A^{\mathscr{H}} \right|_\Gamma,$$

which can be solved to give

$$\gamma^{-1}\big|_\Gamma = \eta(t)\, e^{\mathscr{C}\phi/k}\, \bar{\gamma}^{-1}\big|_\Gamma , \tag{3.103}$$

where $\eta(t)$ is an arbitrary gauge group element depending only on time, $d\eta = 0$.

In the next step of the procedure, one should plug the solution of (3.103) into the Lagrangian (3.102) and derive the effective deformed particle action. Unfortunately, this procedure is extremely technically difficult for the gauge groups that are more complicated than the 2+1-dimensional Poincaré group (some partial results for the case of 2+1-dimensional (Anti) de Sitter group were presented in [13]). However, it is clear that as a result of the presence of the group element $e^{\mathscr{C}\phi/k}$ in the sewing condition (3.103), the effective particle Lagrangian generically describes momentum space that is a group manifold, not the Lie algebra element, and therefore the resulting theory is necessarily deformed. In the next chapters, we discuss in detail one particular example of such curved momentum space, associated with κ-Poincaré algebra construction.

References

1. Freidel, L., Kowalski-Glikman, J., Smolin, L.: 2+1 gravity and doubly special relativity. Phys. Rev. D **69**, 044001 (2004). arXiv:hep-th/0307085
2. Freidel, L., Starodubtsev, A.: Quantum gravity in terms of topological observables. arXiv:hepth/0501191
3. Smolin, L., Starodubtsev, A.: General relativity with a topological phase: an action principle. arXiv:hep-th/0311163
4. Freidel, L., Kowalski-Glikman, J., Starodubtsev, A.: Particles as Wilson lines of gravitational field. Phys. Rev. D **74**, 084002 (2006). arXiv:gr-qc/0607014
5. Holst, S.: Barbero's Hamiltonian derived from a generalized Hilbert-Palatini action. Phys. Rev. D **53**, 5966 (1996). arXiv:gr-qc/9511026
6. Durka, R., Kowalski-Glikman, J., Szczachor, M.: Supergravity ASA constrained BF theory. Phys. Rev. D **81**, 045022 (2010). arXiv:0912.1095 [hep-th]
7. Kowalski-Glikman, J., Starodubtsev, A.: Can we see gravitational collapse in (quantum) gravity perturbation theory? arXiv:gr-qc/0612093
8. Mathisson, M.: Acta. Phys. Polon. **6**, 225 (1937); Papapetrou, A.: Spinning test particles in general relativity. 1. Proc. Roy. Soc. Lond. A **209**, 248 (1951)
9. Kiefer, C.: Quantum Gravity. Oxford University Press (2012)
10. Nakahara, M.: Geometry, Topology and Physics. IOP Publishing (2003)
11. Kowalski-Glikman, J.: Note on gravity, entropy, and BF topologicalfield theory. Phys. Rev. D **81**, 084038 (2010). https://doi.org/10.1103/PhysRevD.81.084038, arXiv:1002.1035 [hep-th]
12. Kowalski-Glikman, J., Starodubtsev, A.: Effective particle kinematics from quantum gravity. Phys. Rev. D **78**, 084039 (2008). https://doi.org/10.1103/PhysRevD.78.084039, arXiv:0808.2613 [gr-qc]
13. Trześniewski, T.: Effective Chern-Simons actions of particles coupled to 3D gravity. Nucl. Phys. B **928**, 448–466 (2018). https://doi.org/10.1016/j.nuclphysb.2018.01.023. arXiv:1706.01375 [hep-th]

Part II
Deformed Particles and Their Symmetries

Deformed Classical Particles: Phase Space and Kinematics

4

In Part I of these notes, we showed how gravity in 2+1 dimensions deforms particle kinematics. It turned out that the original dynamical degrees of freedom of the particle merge with the 'would be gauge' degrees of freedom of the gravitational field making the particle's kinematics effectively deformed. One could say that the deformed particle Lagrangian describes the particle along with the back reaction of its own gravitational field. Another way of thinking about the deformation is that, effectively, the originally flat momentum space becomes, as a result of incorporating this back reaction, a curved manifold, which, in the cases considered above, became a group manifold of the group SO(2, 1) or AN(2).

In this chapter, we will first introduce a group-theoretic formulation of the phase space of particles with a Lie group momentum space based on the theory of Poisson–Lie groups [1]. This will naturally lead to the description of kinematical symmetries in terms of Hopf algebras, the subject of the next chapter. In the second part of this chapter, we illustrate an alternative Lagrangian formulation of the kinematics of deformed particles which, in principle, can be generalized to more curved momentum spaces.

4.1   From Symplectic Manifolds to Poisson–Lie Groups

We begin our discussion with a review of equivalent formulations of the phase space of a classical particle. We start from the conventional picture in terms of the cotangent bundle of the particle's configuration space equipped with a symplectic form. Such symplectic structure together with the Hamiltonian determines the dynamics of the system. We recall how this familiar picture can be recast in the more abstract language of Poisson–Lie groups and associated r-matrices which allow a rather straightforward generalization to phase spaces which *do not* possess the structure of

© Springer-Verlag GmbH Germany, part of Springer Nature 2021
M. Arzano and J. Kowalski-Glikman, *Deformations of Spacetime Symmetries*,
Lecture Notes in Physics 986,
https://doi.org/10.1007/978-3-662-63097-6_4

cotangent bundles and, in particular, to phase spaces in which momenta belong to a non-abelian Lie group. Below we assume that the reader has some basic acquaintance with differential geometry and we do not define many notions we use; the reader could consult any differential geometry textbook, for example, [2].

4.1.1   From Symplectic Manifolds to Poisson Manifolds

The possible states of a classical mechanical system are described by the points of a phase space given by a *symplectic manifold*: an even-dimensional smooth manifold Γ equipped with a non-degenerate closed two-form ω. In most cases, Γ is the cotangent bundle of the configuration space M: $\Gamma = T^*M$. Smooth functions on such phase space, i.e., elements of $C^\infty(\Gamma)$, are the observables of such system. The dynamics are determined by a function on Γ, the Hamiltonian, and the evolution of the system is described by an integral curve of the Hamiltonian vector field X_H on Γ determined by Hamilton equations [3].

The action of the Hamiltonian vector field X_H on functions in $C^\infty(\Gamma)$ can be given in terms of the Poisson bracket, an antisymmetric map $\{\cdot, \cdot\} : C^\infty(\Gamma) \times C^\infty(\Gamma) \to C^\infty(\Gamma)$ which, like a Lie bracket, satisfies the Jacobi identity and the Leibniz rule. If the Poisson bracket is non-degenerate (there is no point in Γ in which $\{f, h\} = 0$ for any $f, h \in C^\infty(\Gamma)$), the Poisson structure is *symplectic*. The Hamiltonian vector field X_f can be defined, for any function $f \in C^\infty(\Gamma)$, via the relation

$$\omega(\cdot, X_f) = df, \tag{4.1}$$

from which, the Poisson bracket can be written in terms of the symplectic form as

$$\{f, g\} = -\omega(X_f, X_g) = -X_f(g). \tag{4.2}$$

The properties which characterize the Poisson bracket are determined by the symplectic form ω, in particular, the antisymmetry is given by the two-form nature of ω, the Jacobi identity corresponds to $d\omega = 0$, and the exterior differentiation accounts for the Leibniz-like property.

Formally we can generalize the mathematical description of the phase space from that of a symplectic manifold to a *Poisson manifold*. This is nothing but a pair $(\Gamma, \{\,,\,\})$ where Γ is a smooth manifold and $\{\,,\,\}$ a Poisson bracket with the properties specified above. The Poisson bracket can be expressed in terms of a skew-symmetric rank-two tensor $w \in T\Gamma \otimes T\Gamma$, an element of the tensor product of two copies of the tangent bundle $T\Gamma$, by means of the *dual pairing* between a vector space and its dual $\langle \cdot, \cdot \rangle : T\Gamma \times T^*\Gamma \to C^\infty(\Gamma)$ and its generalization to tensor fields. Thus, we define the Poisson bracket of the smooth functions $f, g \in C^\infty(\Gamma)$ as

$$\{f, g\} = \langle w, df \otimes dg \rangle, \tag{4.3}$$

i.e., as the pairing of the bivector w with the tensor product of the differentials df and dg. The skew-symmetric tensor w is called *Poisson bivector* and induces a mapping $w(\,,\,\cdot\,) : T^*\Gamma \to T\Gamma$ given by $\langle w, \,\cdot\, \otimes df\rangle = X_f$ which can also be expressed as $X_f = \{\,\cdot\,, f\}$ [4]. The important point to notice is that this map is not necessarily invertible. In the cases when it does not possess an inverse, we do not have a symplectic structure. On the contrary, every symplectic structure is also a Poisson structure.

In this chapter, we gather the necessary tools to describe the phase space and Poisson structure of particles whose momenta belong to a non-abelian Lie group. In general, for phase spaces which are (direct products of) Lie groups, i.e., where both configuration space and momentum space can be curved, there exists a mathematical framework, that of the so-called Poisson–Lie groups, to define Poisson brackets which are compatible with the Lie group structure. The Poisson structure of a Poisson–Lie group, however, is never symplectic. If one renounces to the requirement of compatibility with the group structure, there is still a concise mathematical way to define a symplectic Poisson structure for the Lie group, called the Heisenberg double. Below we provide the basic ingredients needed to determine such symplectic Poisson structure starting from the Lie algebra structure of the configuration and momentum space Lie groups.

4.1.2 Lie Groups as Phase Spaces

Let us start by considering a phase space $\Gamma = T \times G$, given by the Cartesian product of a n-dimensional Lie group configuration space T and a n-dimensional Lie group momentum space G. Of course, in this case, the phase space no longer bears the structure of a cotangent bundle, and it is not obvious how the structures reviewed in the previous section, in particular the Poisson brackets, could be generalized.

We will start by showing how it is possible to define a Poisson structure on the Lie algebras $\mathfrak{t}$ and $\mathfrak{g}$ associated with T and G using their structure constants. We denote the elements of $\mathfrak{t}$ with $\{P_\mu\}$ those of $\mathfrak{g}$ with $\{X^\mu\}$, $\mu = 0, \dots, n-1$. Their Lie brackets are given by

$$[P_\mu, P_\nu] = d^\sigma_{\mu\nu} P_\sigma \qquad \text{and} \qquad [X^\mu, X^\nu] = c^{\mu\nu}_\sigma X^\sigma, \tag{4.4}$$

with $d^\sigma_{\mu\nu}$ and $c^{\mu\nu}_\sigma$ the structure constants of the two algebras. Since elements of the groups T and G describe, respectively, the positions and momenta of a classical system, it is useful to regard $\mathfrak{t}$ and $\mathfrak{g}$ as *dual vector spaces* with a dual pairing defined in terms of the basis elements as

$$\langle P_\mu, X^\nu\rangle = \delta^\nu_\mu. \tag{4.5}$$

Such duality allows us to define Poisson brackets on $\mathfrak{t}$ and $\mathfrak{g}$. To see this, let us consider an element $Y \in \mathfrak{t}$, since $\mathfrak{t}$ is a vector space, the tangent space $T_Y\mathfrak{t} \simeq \mathfrak{t}$ is isomorphic to $\mathfrak{t}$ itself. If we take a smooth function $f \in C^\infty(\mathfrak{t})$, then the differential $(df)_Y : T_Y\mathfrak{t} \to \mathbb{R}$ can be seen as an element of the cotangent space $T^*_Y\mathfrak{t} \simeq \mathfrak{g}$. The

Poisson bracket on $C^\infty(\mathfrak{t})$ is then given in terms of the commutators of $\mathfrak{g}$ by

$$\{f, g\}(Y) \equiv \langle Y, [(df)_Y, (dg)_Y]\rangle . \tag{4.6}$$

In the same way, the Lie brackets on $\mathfrak{t}$ determine Poisson brackets on $C^\infty(\mathfrak{g})$. In particular, let us consider coordinate functions $f = x^\mu$ and $g = x^\nu$ such that $dx^\mu, dx^\nu \in \mathfrak{g}$, it is easy to see that the Lie algebra structure of $\mathfrak{g}$ induces the following Poisson bracket on these functions

$$\{x^\mu, x^\nu\} = c_\sigma^{\mu\nu} x^\sigma , \tag{4.7}$$

and, analogously, the Lie algebra structure on $\mathfrak{t}$ defines a Poisson structure

$$\{p_\mu, p_\nu\} = d_{\mu\nu}^\sigma p_\sigma \tag{4.8}$$

on $C^\infty(\mathfrak{g})$ with p_μ coordinate functions on $\mathfrak{g}$ such that the associated differential coincides with the generators X^μ. Notice that since these brackets are isomorphic to Lie brackets, they automatically satisfy all the required properties for being a Poisson bracket, i.e., skew-symmetry and Jacobi identity. These brackets are known in the literature as Kirillov–Kostant brackets.

Let us now consider two maps $\delta_\mathfrak{t} : \mathfrak{t} \to \mathfrak{t} \otimes \mathfrak{t}$ and $\delta_\mathfrak{g} : \mathfrak{g} \to \mathfrak{g} \otimes \mathfrak{g}$ defined as

$$\delta_\mathfrak{t}(X^\mu) = d_{\alpha\beta}^\mu X^\alpha \otimes X^\beta \quad \text{and} \quad \delta_\mathfrak{g}(P_\mu) = c_\mu^{\alpha\beta} P_\alpha \otimes P_\beta . \tag{4.9}$$

It is easy to see that, through the dual pairing (4.5), the functions $\delta_\mathfrak{t}$ and $\delta_\mathfrak{g}$ determine the Lie brackets of $\mathfrak{g}$ and $\mathfrak{t}$, respectively, via the relations

$$\delta_\mathfrak{t}(X^\mu)(P_\alpha, P_\beta) = \langle X^\mu, [P_\alpha, P_\beta]\rangle , \quad \delta_\mathfrak{g}(P_\mu)(X^\alpha, X^\beta) = \langle P_\mu, [X^\alpha, X^\beta]\rangle , \tag{4.10}$$

and are thus known in the literature as *co-commutators*. Besides their direct relationship with the Poisson structure on the dual space, the co-commutator of a Lie algebra can also be used to introduce a Poisson structure on the associated Lie group. In order to see this, we now review the basics of Poisson structures on a Lie group and in particular, we will derive a formula connecting the Poisson bivector on the group manifold with the co-commutator on its Lie algebra.

4.1.2.1 Poisson Structures on Lie Groups

A Poisson–Lie structure on a Lie group G is a Poisson structure for which the group multiplication $m : G \times G \to G$, denoted as $m(g_1, g_2) = g_1 g_2$, is a *Poisson map*. This property can be expressed in terms of the decomposition of the Poisson bracket for functions on the Cartesian product of spaces as

$$\{f_1 \circ m, f_2 \circ m\}_{G \times G}(g_1, g_2) = \{f_1 \circ m(\cdot, g_2), f_2 \circ m(\cdot, g_2)\}_G(g_1)$$
$$+ \{f_1 \circ m(g_1, \cdot), f_2 \circ m(g_1, \cdot)\}_G(g_2). \tag{4.11}$$

Such expression can be translated into the following condition for the Poisson bracket on one copy of the group $\{\cdot, \cdot\}_G$

$$\{f_1, f_2\}_G(g_1 g_2) = \{f_1 \circ R_{g_2}, f_2 \circ R_{g_2}\}_G(g_1) + \{f_1 \circ L_{g_1}, f_2 \circ L_{g_1}\}_G(g_2),$$
(4.12)

where we used the left and right translation maps

$$L_{g_1}(g_2) = g_1 g_2, \quad R_{g_1}(g_2) = g_2 g_1.$$
(4.13)

A Lie group equipped with a Poisson bracket satisfying such property is known as a *Poisson–Lie group*.

We would now like to rewrite the Poisson bracket in terms of a Poisson bivector, i.e., an element of the tensor product of tangent spaces $T_g G \otimes T_g G$. In doing so, we will be able to describe the infinitesimal version of the Poisson structure on the Lie group and to connect it with the co-commutator defined in the previous subsection. Let us focus on the right translation map R_g. We have the following maps induced by R_{g_2}: the pullback between cotangent spaces $R_{g_2}^* : T_{g_1 g_2}^* G \to T_{g_1}^* G$ and the pushforward between tangent spaces $R_{g_2 *} : T_{g_1} G \to T_{g_1 g_2} G$. Given a tangent vector at $g \in G$, $X \in T_{g_1} G$ and the differential $df \in T_{g_1 g_2} G$, we can write the dual pairing between elements of the dual spaces in terms of these maps as

$$\langle X, R_{g_2}^* df \rangle = \langle R_{g_2 *} X, df \rangle.$$
(4.14)

The analogous relation for the left translation and its induced maps is

$$\langle Y, L_{g_1}^* dh \rangle = \langle L_{g_1 *} Y, dh \rangle,$$
(4.15)

where $Y \in T_{g_1 g_2} G$ and $dh \in T_{g_2}^* G$. The pullback and pushforward of R_g and L_g can be generalized to the tensor product of spaces of any rank using the above relations. The dual pairing between tangent and cotangent spaces can be used to define a Poisson bivector $w_g \in T_g G \otimes T_g G$

$$\{f_1, f_2\}(g) = \langle w_g, df_1 \otimes df_2|_g \rangle,$$
(4.16)

where for notational simplicity we dropped the subscript G on the Poisson bracket. It is rather straightforward to write down the equivalent of the condition (4.12) for the Poisson bivector w_g

$$w_{g_1 g_2} = R_{g_2 *}{}^{\otimes 2}|_{g_1} w_{g_1} + L_{g_1 *}{}^{\otimes 2}|_{g_2} w_{g_2},$$
(4.17)

which states that a Poisson structure for a Lie group G is Poisson–Lie if and only if the value of its Poisson bivector at $g_1 g_2 \in G$ is the sum of the right translate by g_2 of its value at g_1 plus the left translate by g_1 of its value at g_2 (we introduced the rather self-explanatory notation $R_{g_2 *}{}^{\otimes 2}$ and $L_{g_1 *}{}^{\otimes 2}$ for the pushforward of the right and left translation on two copies of the tangent space at a point of the group). Notice that for $g_1 = g_2 = e$, where $e \in G$ is the identity element of the group, we have that $w_e = 0$; therefore, the rank of the Poisson structure is zero at the identity element of the Lie group, hence *the Poisson structure of a Poisson–Lie group is not symplectic.*

Let us stress that it is possible to define other Poisson structures on a Lie group which turn the latter into a symplectic manifold and for which the requirement of being Poisson–Lie is dropped. Since these structures are appealing from a physical standpoint, in what follows we show how the co-commutator at the Lie algebra level and the Poisson bivector defined on the associated Lie group are related. Such a relationship will allow us to introduce a symplectic Poisson structure to the group and to exhibit the differences between Poisson–Lie and Poisson structures.

In order to make contact with the notion of co-commutator, we focus on the right translate of the Poisson bivector w to the identity element of G, denoted by $w^R : G \to \mathfrak{g} \times \mathfrak{g}$, where $\mathfrak{g} = T_e G$ is the Lie algebra of G. First, we note that

$$\langle w_g, (df_1 \otimes df_2)|_g \rangle = \langle w^R(g), R_g^{*\otimes 2}|_e (df_1 \otimes df_2)|_g \rangle,$$
$$= \langle R_{g*}{}^{\otimes 2}|_e w^R(g), (df_1 \otimes df_2)|_g \rangle, \tag{4.18}$$

hence we have for the Poisson bivector,

$$w_g = R_{g*}{}^{\otimes 2} w^R(g). \tag{4.19}$$

In order to obtain an expression involving Lie algebra elements, we express the group element as $g = e^{tX}$ and consider the derivative of the second term in (4.18), using $w^R(e) = 0$, we obtain

$$\frac{d}{dt} \langle w^R(e^{tX}), R_{e^{tX}}^*(df_1 \otimes df_2)|_{e^{tX}} \rangle|_{t=0} = \langle \frac{d}{dt} w^R(e^{tX})\big|_{t=0}, R_{e^{tX}}^*{}^{\otimes 2}(df_1 \otimes df_2)|_{e^{tX}}|_{t=0} \rangle. \tag{4.20}$$

We now *define* the co-commutator in terms of w^R as

$$\frac{d}{dt} w^R(e^{tX})\big|_{t=0} \equiv \delta(X). \tag{4.21}$$

Denoting $\xi_i = df_i|_e$ and equating (4.20) to the derivative of the LHS of (4.18), we can write down the following relation between the co-commutator and the Poisson bracket on the group

$$\langle X, d\{f_1, f_2\}|_e \rangle = \langle \delta(X), \xi_1 \otimes \xi_2 \rangle. \tag{4.22}$$

For Poisson–Lie groups, w^R must comply with a condition equivalent to (4.17) which ensures that the group multiplication on G is a Poisson map. Such condition will translate on a 'local' condition on the co-commutator through (4.21) which we now derive explicitly. Inverting (4.19), we can write

$$w^R(g) = R_{g^{-1}*}{}^{\otimes 2} w_g, \tag{4.23}$$

and acting on (4.17) with the tangent linear map associated to the right action $R_{g_1g_2}^{-1} = R_{(g_1g_2)^{-1}} = R_{g_2^{-1}g_1^{-1}}$, using the identity $R_{g_1}L_{g_2} = L_{g_2}R_{g_1}$, we obtain

$$w^R(g_1g_2) = R_{g_1^{-1}*}^{\otimes 2}L_{g_1*}^{\otimes 2}R_{g_2^{-1}*}^{\otimes 2}w_{g_2} + R_{g_1^{-1}*}^{\otimes 2}w_g \tag{4.24}$$

$$= R_{g_1^{-1}*}^{\otimes 2}L_{g_1*}^{\otimes 2}w^R(g_2) + w^R(g_1). \tag{4.25}$$

The translation $L_gR_{g^{-1}}g' = gg'g^{-1} = \mathrm{Ad}_g g'$ is the adjoint action of g on g' in G. Then, denoting the action of the tangent linear maps as $L_{g_1*}R_{g_1^{-1}*} = \mathrm{Ad}_{g_1}$, we can write the above relation as

$$w^R(g_1g_2) = \mathrm{Ad}_{g_1}^{\otimes 2}w^R(g_2) + w^R(g_1), \tag{4.26}$$

which is the required condition on w^R which ensures that the Poisson structure is compatible with group multiplication. Let us differentiate (4.26) in order to derive a condition for the co-commutator $\delta(X)$. We start by noticing that from the definition (4.21), it follows[1] that $\delta(-X) = -\delta(X)$. Next, we look at the co-commutator of $[X, Y]$

$$\delta([X, Y]) = \delta\left(\frac{d}{ds}\frac{d}{dt}(e^{sX}e^{tY}e^{-sX})|_{s,t=0}\right)$$

$$= \frac{d}{ds}\frac{d}{dt}w^R(e^{sX}e^{tY}e^{-sX})|_{s,t=0}, \tag{4.27}$$

taking into account the identity

$$[X, Y] = \frac{d}{dt}\frac{d}{ds}\left(\mathrm{Ad}_{e^{sX}}e^{tY}\right)\Big|_{t,s=0} = \frac{d}{ds}\left(\mathrm{Ad}_{e^{sX}}\right)\Big|_{s=0}\frac{d}{dt}e^{tY}\Big|_{t=0}$$

$$= \frac{d}{ds}\left(\mathrm{Ad}_{e^{sX}}\right)\Big|_{s=0}Y, \tag{4.28}$$

and applying the condition (4.26) for w^R twice, we get

$$\delta([X, Y]) = \frac{d}{ds}\frac{d}{dt}\left[w^R(e^{sS}) + \mathrm{Ad}_{e^{sX}}^{\otimes 2}w^R(e^{tY}e^{-sX})\right]_{s,t=0}$$

$$= \frac{d}{ds}\frac{d}{dt}\left[w^R(e^{sS}) + \mathrm{Ad}_{e^{sX}}^{\otimes 2}\left(w^R(e^{tY}) + \mathrm{Ad}_{e^{tY}}^{\otimes 2}w^R(e^{-sX})\right)\right]_{s,t=0}$$

$$= \frac{d}{ds}\left(\mathrm{Ad}_{e^{sX}}^{\otimes 2}\right)\frac{d}{dt}w^R(e^{tY})|_{t=0} + \frac{d}{ds}\left(\mathrm{Ad}_{e^{sX}}^{\otimes 2}\right)\frac{d}{dt}\left(\mathrm{Ad}_{e^{tY}}^{\otimes 2}\right)w^R(e^{-sX})|_{s,t=0}$$

$$+ \mathrm{Ad}_{e^{sX}}^{\otimes 2}\frac{d}{dt}\left(\mathrm{Ad}_{e^{tY}}^{\otimes 2}\right)\frac{d}{ds}w^R(e^{-sX})|_{s,t=0}, \tag{4.29}$$

---

[1] The property is easily verified by taking the derivative in the direction of X of $w^R(e) = 0$.

where we used $w^R(e) = 0$, $\mathrm{Ad}_e^{\otimes 2} = \mathbb{1} \otimes \mathbb{1}$. Introducing the notation

$$X.\delta(Y) = (\mathrm{ad}_X \otimes \mathbb{1} + \mathbb{1} \otimes \mathrm{ad}_X)\delta(Y)\,, \tag{4.30}$$

where ad_X denotes the adjoint action of Lie algebra generators on other elements of the algebra, we can rewrite (4.29) as the following equation for the co-commutator

$$\delta([X, Y]) = X.\delta(Y) - Y.\delta(X)\,. \tag{4.31}$$

Equation (4.31) above is known in the mathematical literature as the *co-cycle condition* [4]. A Lie algebra equipped with a co-commutator satisfying the co-cycle condition is called a *Lie bi-algebra*. In what follows, we will show that the Lie bi-algebra structure can give rise to Poisson structures on the Lie group which are not Poisson-Lie but are symplectic. The requirement (4.31) for the Lie bi-algebra structure will turn out to be crucial in order to have a proper Poisson structure on the corresponding group. Our goal in the next sections will be to look for co-commutators which satisfy the co-cycle condition and which we can use to construct a Poisson structure on a phase space with group-valued momenta.

4.1.2.2 Poisson Structures and the *r*-Matrix

One way to introduce a co-commutator which automatically satisfies the co-cycle condition is to consider one of the form

$$\delta(X) \equiv X.r = (\mathrm{ad}_X \otimes \mathbb{1} + \mathbb{1} \otimes \mathrm{ad}_X)r\,, \tag{4.32}$$

where r is a generic element of the tensor product $\mathfrak{g} \otimes \mathfrak{g} = \otimes^2\mathfrak{g}$ called the *r-matrix*. In order for δ to be a genuine co-commutator, the r-matrix must satisfy the following two conditions:

1. The symmetric part of r, $r_+ = \frac{1}{2}\left(r^{ij} + r^{ji}\right) X_i \otimes X_j$ with $r = r^{ij} X_i \otimes X_j$ and $\{X_i\}$ a basis of $\mathfrak{g}$, is an ad-invariant element of $\otimes^2\mathfrak{g}$.
2. The combination $\left[[r, r]\right] = [r_{12}, r_{13}] + [r_{12}, r_{23}] + [r_{13}, r_{23}]$, where

$$r_{12} = r^{ij}\, X_i \otimes X_j \otimes \mathbb{1},$$
$$r_{13} = r^{ij}\, X_i \otimes \mathbb{1} \otimes X_j,$$
$$r_{23} = r^{ij}\, \mathbb{1} \otimes X_i \otimes X_j, \tag{4.33}$$

and

$$[r_{12}, r_{13}] = r^{ij} r^{kl} [X_i, X_k] \otimes X_j \otimes X_l,$$
$$[r_{12}, r_{23}] = r^{ij} r^{kl} X_i \otimes [X_j, X_k] \otimes X_l,$$
$$[r_{13}, r_{23}] = r^{ij} r^{kl} X_i \otimes X_k \otimes [X_j, X_l], \tag{4.34}$$

known as the *Schouten bracket*, is an ad-invariant element of $\otimes^3\mathfrak{g}$.

The first condition is directly related to the skew-symmetry of the Poisson bracket and of the Lie brackets defined by the co-commutator δ. Explicitly, it requires that

$$(\text{ad}_X \otimes \mathbb{1} + \mathbb{1} \otimes \text{ad}_X)r_+ = 0, \tag{4.35}$$

and such equation is trivially satisfied if $r_+ = 0$, i.e., adopting a skew-symmetric r-matrix. The second condition ensures that such brackets satisfy the Jacobi identity. It is also clear that the simplest way of satisfy such condition is if $\big[[r, r]\big] = 0$. This equation is known as the *Classical Yang–Baxter Equation* (CYBE) and its solution is called the *classical r-matrix*. The condition which imposes the Schouten bracket $\big[[r, r]\big]$ to be ad-invariant

$$X^i.\big[[r, r]\big] = 0 \quad \forall X^i \in \mathfrak{g}, \tag{4.36}$$

where

$$X^i.\big[[r, r]\big] \equiv \big(\text{ad}_{X^i} \otimes \mathbb{1} \otimes \mathbb{1} + \mathbb{1} \otimes \text{ad}_{X^i} \otimes \mathbb{1} + \mathbb{1} \otimes \mathbb{1} \otimes \text{ad}_{X^i}\big)\big[[r, r]\big], \tag{4.37}$$

is known as the *modified Classical Yang–Baxter Equation* (mCYBE). The most important point for us is that the co-commutator, as defined by (4.32), can be integrated to a Poisson structure on G. One possibility is to insist on the requirement of compatibility with group multiplication thus obtaining a Poisson–Lie structure on G. It is important to stress, however, that other Poisson structures can be associated to the same co-commutator which are not Poisson–Lie but which, for example, are symplectic and thus good candidates for describing the Poisson brackets of a deformed phase space. These last structures are the ones we are interested in.

Before moving on, we will describe in some detail the Poisson bivectors associated to a r-matrix. A possible choice of Poisson bivector associated to a given r-matrix is such that its right translate to the identity element of G is given by

$$w^R(g) = \text{Ad}_g^{\otimes 2} r - r. \tag{4.38}$$

Indeed, writing an element of G as $g = e^{tX}$ is a straightforward calculation to check that (4.38) has the 'correct' derivative

$$\delta(X) = \frac{d}{dt} w^R(e^{tX})\Big|_{t=0}$$
$$= \left[\frac{d}{dt}\big(\text{Ad}_{e^{tX}}\big) \otimes \text{Ad}_{e^{tX}} + \text{Ad}_{e^{tX}} \otimes \frac{d}{dt}\big(\text{Ad}_{e^{tX}}\big)\right]_{t=0} r$$
$$= (\text{ad}_X \otimes \mathbb{1} + \mathbb{1} \otimes \text{ad}_X)\, r = X.r. \tag{4.39}$$

It is also easily checked that (4.38) satisfies the co-cycle property (4.26).

It is possible to define a different Poisson structure on G, starting from the same r-matrix, which also satisfies the above properties. This structure is determined by the Poisson bivector

$$w^R(g) = \text{Ad}_g^{\otimes 2} r + r^*, \tag{4.40}$$

where r^* is minus the transpose of r, that is, if $r = r^{ij} X_i \otimes X_j$ then $r^* = -r^{ji} X_i \otimes X_j$, $(r^* = -r^t)$. As we will see, this Poisson bivector *does not* give rise to a Poisson–Lie structure but it allows to define a symplectic Poisson structure on the group manifold since it is not necessarily degenerate on the group identity.

It is useful to write down the explicit form of the Poisson brackets on the group in terms of the r-matrix. The Poisson bracket associated to (4.38) can be obtained by writing (4.16) in terms of the right translate of w_g to the identity element, see Equation (4.18)

$$
\begin{aligned}
\{f_1, f_2\}(g) &= \langle w^R(g), R_g^{*\otimes 2} (df_1 \otimes df_2)|_g \rangle \\
&= \langle (\mathrm{Ad}_g^{\otimes 2} r - r, R_g^{*\otimes 2} (df_1 \otimes df_2)|_g \rangle \\
&= \langle L_{g*} R_{g^{-1}*}^{\otimes 2} r, R_g^{*\otimes 2} (df_1 \otimes df_2)|_g \rangle \\
&\qquad - \langle r, R_g^{*\otimes 2} (df_1 \otimes df_2)|_g \rangle \\
&= \langle r, L_g^{*\otimes 2} (df_1 \otimes df_2)_g \rangle - \langle r, R_g^{*\otimes 2} (df_1 \otimes df_2)|_g \rangle .
\end{aligned}
\tag{4.41}
$$

where in the third line, we used the equality $L_{g_1*} R_{g_1^{-1}*} = \mathrm{Ad}_{g_1}$. Thus, we have

$$
\{f_1, f_2\}(g) = \langle r, L_g^{*\otimes 2} (df_1 \otimes df_2)_g \rangle - \langle r, R_g^{*\otimes 2} (df_1 \otimes df_2)|_g \rangle .
$$

The corresponding expression for the bivector (4.40) is given by

$$
\{f_1, f_2\}(g) = \langle r, L_g^{*\otimes 2}(df_1 \otimes df_2)|_g \rangle + \langle r^*, R_g^{*\otimes 2}(df_1 \otimes df_2)|_g \rangle .
\tag{4.42}
$$

In the next section, we will apply the tools illustrated so far to Cartesian product Lie groups of the type $T \times G$. Imposing a duality relation between the Lie algebras $\mathfrak{t}$ and $\mathfrak{g}$, the group manifold $\Gamma = T \times G$ can be seen as a deformation of ordinary phase spaces with group manifold configuration space T *and* momentum space G.

4.1.3 Deforming Phase Spaces: The Classical Doubles

In order to define a Poisson structure on the Lie group phase space $\Gamma = T \times G$, we will look for an 'exponentiated' version of the co-commutators defined on the Lie (bi)-algebra $\mathfrak{t} \oplus \mathfrak{g}$ associated to the group Γ. The starting point, of course, will be to define a Lie algebra structure on the vector space $\mathfrak{t} \oplus \mathfrak{g}$ compatible with the duality relation (4.5) between $\mathfrak{t}$ and $\mathfrak{g}$. Such relation is expressed in terms of the natural inner product on $\mathfrak{t} \oplus \mathfrak{g}$ given by

$$
(P_\mu, P_\nu) = 0, \qquad (X^\mu, X^\nu) = 0, \qquad \text{and} \qquad (P_\mu, X^\nu) = \langle P_\mu, X^\nu \rangle .
\tag{4.43}
$$

We want to define on $\mathfrak{t} \oplus \mathfrak{g}$ a Lie bracket such that the inner product above is *invariant* under the adjoint action of the elements of $\mathfrak{t} \oplus \mathfrak{g}$, that is,

$$
([Z_A, Z_B], Z_C) = (Z_A, [Z_B, Z_C]),
\tag{4.44}
$$

where $Z_A = \{P_\mu, X^\mu\}$, $A = 1, \ldots 2n$. The following Lie brackets

$$[P_\mu, P_\nu] = d^\sigma_{\mu\nu} P_\sigma, \qquad [X^\mu, X^\nu] = c^{\mu\nu}_\sigma X^\sigma, \qquad \text{and} \qquad [P_\mu, X^\nu] = c^{\nu\sigma}_\mu P_\sigma - d^\nu_{\mu\sigma} X^\sigma \tag{4.45}$$

comply with such requirement as it can be easily verified. However, in order to show that these brackets turn $\mathfrak{t} \oplus \mathfrak{g}$ into a Lie algebra, we must ensure that they satisfy the Jacobi identity. It turns out that the brackets (4.45) on $\mathfrak{t} \oplus \mathfrak{g}$ satisfy the Jacobi identity *if and only if* the co-commutator $\delta_\mathfrak{t}$ on the Lie algebra $\mathfrak{t}$ satisfies the co-cycle condition (4.31) [4]. As we discussed in Sect. 4.1.2, the Lie algebra structure on $\mathfrak{t}$ defines a co-commutator $\delta_\mathfrak{g}$ on the dual 'momentum' Lie algebra $\mathfrak{t}$ via

$$\langle P, [X_i, X_j] \rangle = \langle \delta_\mathfrak{g}(P), X_i \otimes X_j \rangle. \tag{4.46}$$

Thus, we see that given the Lie algebra structure (4.45) on $\mathfrak{t} \oplus \mathfrak{g}$ and the pairing through the inner product (4.43), a Lie bi-algebra structure on $\mathfrak{t}$ (and by duality on $\mathfrak{g}$) is naturally induced and it can be shown to be *unique* [4].

We want to define now a Lie bi-algebra structure on the whole direct sum Lie algebra $\mathscr{D} = \mathfrak{t} \oplus \mathfrak{g}$, i.e., define a co-commutator $\delta_\mathscr{D}$ which reproduces $\delta_\mathfrak{t}$ and $\delta_\mathfrak{g}$ as given in Eq. (4.9) when restricted, respectively, to $\mathfrak{t}$ and $\mathfrak{g}$. It turns out that there is a canonical way of defining such co-commutator in terms of the r-matrix[2] belonging to $\mathfrak{t} \otimes \mathfrak{g}$

$$r = P_\mu \otimes X^\mu, \tag{4.47}$$

as

$$\delta_\mathscr{D}(Z_A) = Z_A . r = \left(\text{ad}_{Z_A} \otimes \mathbb{1} + \mathbb{1} \otimes \text{ad}_{Z_A} \right) r. \tag{4.48}$$

It is easily verified that such co-commutator reduces to (4.9), when $Z_A = X^\mu$ or $Z_A = P_\mu$. It can be also proved [4] that (4.48) defines a genuine Lie bi-algebra structure on $\mathscr{D}$, i.e., that $\delta_\mathscr{D}(Z_A)$ complies with the properties listed in Sect. 4.1.2.

We can now use the r-matrix (4.47) to define a Poisson structure along the lines illustrated in the previous section. There we saw that, given an appropriate r-matrix, there are two possible choices of Poisson bivector which can be used to define a Poisson structure on $\Gamma = T \times G$. Let us consider two functions on Γ, $f_1, f_2 \in C^\infty(T \times G)$, then the Poisson bracket evaluated in $h \in T \times G$ is given by $\{f_1, f_2\}(h) = \langle w_h, (df_1 \otimes df_2)_h \rangle$. The two Poisson brackets corresponding to (4.42) and (4.42) can be written in terms of the r-matrix as

$$\{f_1, f_2\} = -r^{AB} \left(Z^R_A f_1 \, Z^R_B f_2 \pm Z^L_A f_1 \, Z^L_B f_2 \right), \tag{4.49}$$

where $Z^L_A = L_{h*} Z_A$ and $Z^R_B = R_{h*} Z_B$ are the left and right translates of $Z_A \in \mathfrak{t} \oplus \mathfrak{g}$. If Γ is a matrix group, then (4.49) can be written as

$$\{\gamma_{ij}, \gamma_{kl}\} = -\sum_{a,b} (r_{iakb} \gamma_{aj} \gamma_{bl} \pm r_{ajbl} \gamma_{ia} \gamma_{kb}), \tag{4.50}$$

---

[2]Sometimes, the r-matrix is written directly as the skew-symmetric part $r_- = \frac{1}{2} \left(P_\mu \otimes X^\mu - X^\mu \otimes P_\mu \right) \equiv P_\mu \wedge X^\mu$, which obviously yields the same Lie bi-algebra structure.

where the components γ_{ij} of $\gamma \in \Gamma$ can be understood as coordinate functions for the group and thus $r_{iakb} \equiv (r^{AB} Z_A \otimes Z_B)_{iakb}$, and $Z_A^L(\gamma_{ij}) = (\gamma Z_A)_{ij}$, $Z_A^R(\gamma_{ij}) = (Z_A \gamma)_{ij}$. The Poisson brackets can be expressed compactly as

$$\{\gamma \overset{\otimes}{,} \gamma\} = -[r, \gamma \otimes \gamma]_{\pm} \tag{4.51}$$

with the plus subscript denoting the anticommutator and the minus for the commutator. The bracket given by the commutator equips $\mathscr{D}$ with the structure of *Drinfeld double* and the one with the anticommutator the structure of the *Heisenberg double*. We would like to stress the fundamental difference between these two structures. On one hand, the Drinfeld structure (4.38) satisfies the compatibility condition (4.17) for having a Poisson–Lie group, but the Poisson structure is always degenerated at the identity element $e \in G$, hence the structure is not symplectic.[3] On the other hand, in the case of the Heisenberg double, one renounces to the requirement of compatibility between the group multiplication and the Poisson structure favouring the possibility of a global symplectic structure [6].

In what follows, we will focus on explicit examples of phase spaces and their Poisson structures making use of the tools developed so far. We will start from the phase space of an ordinary massive relativistic particle in the Minkwowski space and then proceed to consider *deformed phase spaces* in three and four spacetime dimensions focussing on the momentum Lie groups encountered in Chap. 2.

4.2   Relativistic Spinless Particle: Flat Momentum Space

In this section, we start by describing the (undeformed) phase space of a relativistic spinless particle using the tools developed in the previous section. Even though what follows can appear as a purely academic exercise, it will actually serve as a starting point for introducing the deformations of momentum space which we will develop in the next sections.

The configuration space of a spinless relativistic particle in $n + 1$-dimensional Minkowski spacetime can be identified with the abelian group of translations, $\mathscr{T} \simeq \mathbb{R}^{n,1}$. At any point in the configuration space $x \in \mathscr{T}$, the cotangent (momentum) space is $T_x^* \mathbb{R}^{n,1} = \mathbb{R}^{n,1*}$, where $\mathbb{R}^{n,1*}$ stands for the dual, as a vector space, to $\mathbb{R}^{n,1}$. Following a 'geometric' approach, at this point, one would introduce a symplectic structure and with it a Poisson bracket for the space of functions on the cotangent bundle $T^* \mathscr{T}$. Here, we will follow an algebraic approach along the lines discussed in the previous section. Our phase space manifold Γ is given by

$$\Gamma = \mathscr{T} \times \mathscr{T}^* \simeq \mathbb{R}^{n,1} \times \mathbb{R}^{n,1*}. \tag{4.52}$$

We denote the Lie algebra associated to each component of Γ as $\mathfrak{t}$ and $\mathfrak{t}^*$ for $\mathscr{T}$ and $\mathscr{T}^*$, respectively, and the coordinates for $\mathscr{T}$ as x^μ and p_μ for $\mathscr{T}^*$, with

---

[3]It is possible, however, to foliate the group manifold in a set of *symplectic leaves* with the Poisson structure of the manifold restricted to each leaf [5].

$\mu = 0, \ldots, n$. For the Lie algebra t, we denote the basis elements as $\{P_\mu\}$, whereas for t^*, we use $\{X^\mu\}$. The (trivial) Lie brackets on the Lie algebras t and t^* are

$$[P_\mu, P_\nu] = 0 \quad \text{and} \quad [X^\mu, X^\nu] = 0, \tag{4.53}$$

for all μ, ν. We will see that the coordinate bases for the Lie algebras related to the groups coordinates are

$$P_\mu = \frac{\partial}{\partial x^\mu} \quad \text{and} \quad X^\mu = -\frac{\partial}{\partial p_\mu}, \tag{4.54}$$

where $X^\mu = \eta^{\mu\nu} X_\nu$, $\eta^{\mu\nu} = \text{diag}(+, -, \ldots, -)$. Taking into account that t and t^* are dual as vector spaces $\langle P_\mu, X^\nu \rangle = \delta_\mu^\nu$, we can define an inner product for $t \oplus t^*$ extending the dual pairing as follows:

$$(P_\mu, P_\nu) = (X^\mu, X^\nu) = 0 \quad \text{and} \quad (P_\mu, X^\nu) = \langle P_\mu, X^\nu \rangle = \delta_\mu^\nu. \tag{4.55}$$

As we saw in the previous section, we can define a Lie bracket for $t \oplus t^*$ asking that the inner product of the direct sum is ad-invariant. The 'mixed' commutator vanishes for any two members of $t \oplus t^*$, thus we obtain an Abelian Lie algebra

$$[P_\mu, P_\nu] = 0, \quad [X^\mu, X^\nu] = 0, \quad \text{and} \quad [P_\mu, X^\nu] = 0. \tag{4.56}$$

Therefore, for t and t^*, we have trivial co-commutators

$$\delta_t(P_\mu) = 0 \quad \text{and} \quad \delta_{t^*}(X^\mu) = 0, \tag{4.57}$$

in accordance with the two first expressions in (4.56).

At this point, we only have a Lie algebra structure for the direct sum of vector spaces $t \oplus t^*$. We would like to introduce a Lie bi-algebra structure on $t \oplus t^*$ in order to obtain the double of the Lie algebra of translations which we denote with $\mathscr{D}(t)$. Since we are dealing with an abelian Lie algebra, the Lie bi-algebra structure will have a trivial co-commutator. Nevertheless, we want to carry out this construction explicitly since the non-trivial cases which we will discuss later can be interpreted as *deformations* of this standard abelian case in which momentum space becomes a non-abelian Lie group.

The co-commutator for the double $\mathscr{D}(t)$ can be obtained from (4.48) by choosing an r-matrix which fulfils the conditions of ad-invariance of its symmetric part and the (modified) Yang–Baxter equation. Since the Lie algebra $t \oplus t^*$ is trivial, any r-matrix satisfies the Lie bi-algebra conditions and has a trivial co-commutator. The canonical Poisson bracket can be obtained from the following antisymmetric r-matrix:

$$r = \frac{1}{2}\left(P_\mu \otimes X^\mu - X^\mu \otimes P_\mu\right) \equiv P_\mu \wedge X^\mu. \tag{4.58}$$

Grouping the set of generators for t and t* as $Z^A = (P_\mu, X^\mu)$ where $A = 1, \dots, 2(n + 1)$, we see that

$$\delta_{\mathscr{D}(t)}(Z^A) = (\text{ad}_{Z^A} \otimes \mathbb{1} + \mathbb{1} \otimes \text{ad}_{Z^A})r = 0 \tag{4.59}$$

for all $Z^A \in t \oplus t^*$, i.e., we have Lie bi-algebra structure turning $t \oplus t^*$ in a classical double.

We can now derive the Poisson bracket for the Poisson–Lie group whose Lie algebra is given by $t \oplus t^*$ with the trivial commutators (4.56). From the canonical r-matrix (4.58) and choosing the plus sign in (4.49), we see that

$$\begin{aligned}
\{f_1, f_2\} &= -2r^{AB} Z_A f_1 Z_B f_2 \\
&= -\delta_\nu^\mu \left(P_\mu f_1 X^\nu f_2 - X^\nu f_1 P_\mu f_2 \right).
\end{aligned} \tag{4.60}$$

Using the coordinate basis (4.54), we obtain

$$\{f_1, f_2\} = \frac{\partial f_1}{\partial x^\mu} \frac{\partial f_2}{\partial p_\mu} - \frac{\partial f_1}{\partial p_\mu} \frac{\partial f_2}{\partial x^\mu}, \tag{4.61}$$

and the Poisson brackets for the coordinate functions on the phase space are

$$\begin{array}{llll}
\{x^0, x^a\} = 0 & \{x^a, x^b\} = 0 & \{p_0, p_a\} = 0 & \{p_a, p_b\} = 0 \\
\{x^0, p_0\} = 1 & \{x^a, p_0\} = 0 & \{x^0, p_a\} = 0 & \{x^a, p_b\} = \delta_b^a,
\end{array} \tag{4.62}$$

where $a, b = 1, \dots, n$. This Poisson structure is also symplectic as the Poisson bivector is non-degenerate. Notice that if we had chosen the minus sign in (4.49), i.e., the Drinfeld double structure, we would have obtained a trivial Poisson–Lie structure! Thus, the Heisenberg double structure, in flat momentum space, can be used to define the canonical Poisson brackets of standard textbook classical mechanics of a relativistic point particle.

4.3   Deforming Momentum Space to the AN(n) Group

As the first example of non-abelian momentum space, we will consider the Lie group AN(n), the n-dimensional generalization of the AN(2) group already discussed in Chap. 2. In this section, we will be show that starting from *minimal* ingredients, namely the structure constants of the Lie algebra $\mathfrak{an}(n)$ of such momentum space Lie group, we can reproduce the Poisson brackets found in Sect. 2.6. Such phase space provides a description of the kinematics of a classical κ-deformed relativistic particle.

Our starting point will be the manifold $\Gamma = T \times \text{AN}(n)$ where $T = \mathbb{R}^{n,1}$ that is the ordinary $n + 1$-dimensional Minkowski configuration space. The usual flat momentum space, however, is now replaced by the group manifold AN(n). In all examples that will follow, we will restrict to models with flat configuration space $T = \mathbb{R}^{n,1}$.

Let us write down the Lie algebras of both components of the Cartesian product group Γ. Denoting with $\{P_\mu\}$ the basis of t and with $\{\tilde{X}^\mu\}$, $\mu = 0, \ldots, n$ the basis of $\mathfrak{an}(n)$, we have

$$[P_\mu, P_\nu] = 0 \quad \text{and} \quad [\tilde{X}^\mu, \tilde{X}^\nu] = -\frac{1}{\kappa}\left(\tilde{X}^\mu \delta_0^\nu - \tilde{X}^\nu \delta_0^\mu\right). \tag{4.63}$$

We immediately see that in the limit $\kappa \to \infty$, we recover the undeformed case of a ordinary relativistic particle reviewed in the previous section. The algebra $\mathfrak{an}(n)$ is usually expressed as

$$[\tilde{X}^0, \tilde{X}^a] = \frac{1}{\kappa}\tilde{X}^a \quad [\tilde{X}^a, \tilde{X}^b] = 0, \tag{4.64}$$

where $a, b = 1, \ldots, n$. The two vector spaces t and $\mathfrak{an}(n)$ are dual with respect to the inner product $\langle P_\mu, \tilde{X}^\nu \rangle = \delta_\mu^\nu$.

The crucial point which distinguishes the case of the ordinary relativistic particle from the κ-deformed model under consideration is that the non-abelian Lie algebra structure of $\mathfrak{an}(n)$ reflects on a non-trivial co-commutator for t

$$\delta_t(P_\mu) = -\frac{1}{\kappa}\left(P_\mu \otimes P_0 - P_0 \otimes P_\mu\right) = \frac{2}{\kappa}\left(P_0 \wedge P_\mu\right), \tag{4.65}$$

while for $\mathfrak{an}(n)$, we have

$$\delta_{\mathfrak{an}}(\tilde{X}^\mu) = 0. \tag{4.66}$$

The direct sum of Lie algebras $\mathfrak{t} \oplus \mathfrak{an}(n)$ can be equipped with an inner product invariant under the action of t and $\mathfrak{an}(n)$ (see (4.44))

$$(P_\mu, P_\nu) = 0, \quad (\tilde{X}^\mu, \tilde{X}^\nu) = 0, \quad \text{and} \quad (P_\mu, \tilde{X}^\nu) = \delta_\mu^\nu. \tag{4.67}$$

Such product can be used to derive commutators defining a Lie algebra structure on $\mathfrak{t} \oplus \mathfrak{an}(n)$, these are given by (4.63) together with

$$[P_\mu, \tilde{X}^\nu] = -\frac{1}{\kappa}\left(\delta_\mu^\nu P_0 - \delta_0^\nu P_\mu\right), \tag{4.68}$$

which written explicitly read

$$[P_0, \tilde{X}^\mu] = 0 \quad [P_a, \tilde{X}^0] = \frac{1}{\kappa}P_a \quad [P_a, \tilde{X}^b] = -\frac{1}{\kappa}\delta_a^b P_0. \tag{4.69}$$

As we showed in Sect. 4.1.2, in order to define a Poisson structure on Γ, it suffices to introduce an r-matrix which turns $\mathfrak{t} \oplus \mathfrak{an}(n)$ into a Lie bi-algebra. A candidate r-matrix can be obtained from (4.58) simply replacing X^μ with the new generators $\tilde{X}^\mu$

$$r = \frac{1}{2}\left(P_\mu \otimes \tilde{X}^\mu - \tilde{X}^\mu \otimes P_\mu\right) \equiv P_\mu \wedge \tilde{X}^\mu. \tag{4.70}$$

It is easily checked that this skew-symmetric r-matrix (4.70) satisfies the two conditions needed to make $\mathfrak{t} \oplus \mathfrak{an}(n)$ a Lie bi-algebra, that is, $r_+ = 0$ is trivially ad-invariant, and a direct calculation shows that the Schouten bracket satisfies the modified classical Yang–Baxter equation

$$Z_A.[[r, r]] = 0 \quad \forall \, Z_A \in \mathfrak{t} \oplus \mathfrak{an}(n). \tag{4.71}$$

The co-commutator on $\mathfrak{t} \oplus \mathfrak{an}(n)$ is defined by (4.48) and on the generators of $\mathfrak{t}$ and $\mathfrak{an}(n)$ reduces to

$$\delta_{\mathfrak{t}\oplus\mathfrak{an}}(P_\mu) = -\frac{1}{\kappa}\left(P_\mu \otimes P_0 - P_0 \otimes P_\mu\right) = \frac{2}{\kappa} P_0 \wedge P_\mu, \tag{4.72}$$

$$\delta_{\mathfrak{t}\oplus\mathfrak{an}}(\tilde{X}^\mu) = 0. \tag{4.73}$$

It can be verified that the co-commutator satisfies the co-cycle condition

$$\delta_{\tilde{\mathscr{D}}(\mathfrak{t})}([Z_A, Z_B]) = Z_A.\delta_{\tilde{\mathscr{D}}(\mathfrak{t})}(Z_B) - Z_B.\delta_{\tilde{\mathscr{D}}(\mathfrak{t})}(Z_A), \tag{4.74}$$

so the Lie bi-algebra $\mathfrak{t} \oplus \mathfrak{an}(2)$ can be seen as a *classical double* $\tilde{\mathscr{D}}(\mathfrak{t}) \equiv \mathfrak{t} \oplus \mathfrak{an}(n)$.

We can use the structures just described to construct a Poisson structure on the group $\tilde{D}(T) = T \times AN(n)$. We will use a matrix representation for the Lie algebra $\mathfrak{an}(n)$ and we will extend it in order to include in the representation the Lie algebra $\mathfrak{t}$. It is common practice to describe the group $AN(n)$ in embedding coordinates which make clear the identification of the group manifold with half of the $(n + 1)$-de Sitter hyperboloid embedded in $(n + 1) + 1$-Minkowski space. In this case, the generators of the corresponding Lie algebra are given by combinations of the generators of the Lie algebra $\mathfrak{so}(n + 1, 1)$ and are represented by $(n + 2) \times (n + 2)$ matrices [7]. However, in order to include translations and to obtain a matrix representation for the Lie algebra $\mathfrak{t} \oplus \mathfrak{an}(n)$, it will be necessary to work in a different representation. We thus introduce the *adjoint representation* $\mathscr{R}$ for $\mathfrak{an}(n)$ defined by $\mathscr{R} : \mathfrak{an}(n) \to \mathfrak{gl}(\mathfrak{an}(n))$, $\tilde{X} \mapsto \mathrm{ad}_{\tilde{X}}$, where $\mathrm{ad}_{\tilde{X}}(\tilde{Y}) := [\tilde{X}, \tilde{Y}]$ for $\tilde{X}, \tilde{Y} \in \mathfrak{an}(n)$. The generators of the adjoint representation are determined by structure constants of the Lie algebra, $[\tilde{X}^\mu, \tilde{X}^\nu] = c^{\mu\nu}_{\ \ \alpha}\tilde{X}^\alpha$, thus the matrices associated to the representation are given by $[\mathscr{R}(\tilde{X}^\mu)]_\alpha^{\ \beta} = c^{\mu\beta}_{\ \ \alpha}$. It is possible to construct another representation from $\mathscr{R}$ via the matrices $\mathscr{R}^*(\tilde{X}^\mu) = -(\mathscr{R}(\tilde{X}^\mu))^\mathrm{T}$, where the superscript T stands for the transpose of the matrix [8]. We will call this matrix representation the *co-adjoint* representation $\mathscr{R}^*$ of $\mathfrak{an}(n)$.

In what follows, we will work with the co-adjoint representation for $\mathfrak{an}(n)$ since it allows to extend the $(n + 1) \times (n + 1)$-matrix representation of $\mathfrak{an}(n)$ to include the basis of $\mathfrak{t}$ arranged in an extra column resulting in a $(n + 2) \times (n + 2)$-matrix representation for the Lie algebra $\mathfrak{t} \oplus \mathfrak{an}(n)$. Let us write such matrix representation explicitly. The basis for $\mathfrak{an}(n)$ is given by $n + 1$ matrices of size $(n + 1) \times (n + 1)$. The basis of the Lie algebra $\mathfrak{t} \oplus \mathfrak{an}(n)$ can be represented in terms of $(n + 2) \times (n +$

2)-matrices. The matrices corresponding to $\mathfrak{t}$ read

$$P_\mu = \begin{pmatrix} \mathbf{0}_{(n+1)\times(n+1)} & \mathbf{u}_\mu \\ \mathbf{0}^{\mathrm{T}}_{(n+1)} & 0 \end{pmatrix}, \tag{4.75}$$

where $\mathbf{0}_n$ and $\mathbf{0}_{(n+1)}$ are n and $(n+1)$-component zero vectors, respectively, and $\mathbf{u}_\mu = (0,\ldots,1,\ldots,0)$ is a $(n+1)$-component vector with 1 in the μth entry for $\mu = 0,\ldots,n$. The matrices representing the $\mathfrak{an}(n)$ sector are given by

$$\tilde{X}^0 = -\frac{1}{\kappa}\begin{pmatrix} & \mathbf{0}^{\mathrm{T}}_n & \\ \mathbf{0}_{(n+2)} & \mathbb{1}_{n\times n} & \mathbf{0}_{(n+2)} \\ & \mathbf{0}^{\mathrm{T}}_n & \end{pmatrix} \quad \text{and} \quad \tilde{X}^a = \frac{1}{\kappa}\begin{pmatrix} & \mathbf{e}^{\mathrm{T}}_a & \\ \mathbf{0}_{(n+2)} & \mathbf{0}_{n\times n} & \mathbf{0}_{(n+2)} \\ & \mathbf{0}^{\mathrm{T}}_n & \end{pmatrix}, \tag{4.76}$$

where $\mathbf{e}_a = (0,\ldots,1,\ldots,0)$ is a n-component vector with 1 in the ath entry for $a = 1,\ldots,n$. A general group element $d \in \tilde{D}(T) = T \times AN(n)$ can be expressed as $d = t\,g$ where t is a pure translation and g is a pure $AN(n)$ element. The explicit matrix form of the group element is given by

$$d = \begin{pmatrix} \tilde{g} & \mathbf{x}_{n+1} \\ \mathbf{0}^{\mathrm{T}}_{n+1} & 1 \end{pmatrix}, \quad \text{with} \quad g = \begin{pmatrix} \tilde{g} & \mathbf{0}_{n+1} \\ \mathbf{0}^{\mathrm{T}}_{n+1} & 1 \end{pmatrix}, \quad t = \begin{pmatrix} \mathbb{1}_{(n+1)\times(n+1)} & \mathbf{x}_{n+1} \\ \mathbf{0}^{\mathrm{T}}_{n+1} & 1 \end{pmatrix}, \tag{4.77}$$

where $\tilde{g} \in AN(n)$ is a $(n+1) \times (n+1)$ matrix and $\mathbf{x}_{n+1} = (x^0, x^a)$ is a $(n+1)$-component vector with real entries $x^\mu \in \mathbb{R}$ that parametrize the group elements of $T \sim \mathbb{R}^{n,1}, t = e^{x^\mu P_\mu}$. The $AN(n)$ group can be parametrized in different ways using a set of real coordinates $\{p_0,\ldots,p_n\}$ so the matrix group entries are $\tilde{g}_{ij}(p)$.

Before presenting the Poisson brackets for particular parametrizations, we will first write down the results in a general, coordinate-independent, form. Using the matrix representation for $\mathcal{D}(\mathfrak{t})$, the Poisson brackets for the Heisenberg double are determined by

$$\{d \overset{\otimes}{,} d\} = -[r, d \otimes d]_+, \tag{4.78}$$

with the r-matrix given by (4.70). In order to see how the Poisson brackets can be read off the expression above, one should recall that, in a simplified case in which d is a 2×2-matrix, the left hand side of (4.78) would be given, for example, by

$$\{d \overset{\otimes}{,} d\} = \begin{pmatrix} \{d_{11}, d_{11}\} & \{d_{11}, d_{12}\} & \{d_{12}, d_{11}\} & \{d_{12}, d_{12}\} \\ \{d_{11}, d_{21}\} & \{d_{11}, d_{22}\} & \{d_{12}, d_{21}\} & \{d_{12}, d_{22}\} \\ \{d_{21}, d_{11}\} & \{d_{21}, d_{12}\} & \{d_{22}, d_{11}\} & \{d_{22}, d_{12}\} \\ \{d_{21}, d_{21}\} & \{d_{21}, d_{22}\} & \{d_{22}, d_{21}\} & \{d_{22}, d_{22}\} \end{pmatrix}, \tag{4.79}$$

whereas the components in the right hand side of (4.78) are simply those of the product of the matrices $-(r(d \otimes d) - (d \otimes d)r)$. An explicit calculation of (4.78) leads thus to the following Poisson brackets:

$$\{x^0, x^a\} = \frac{1}{\kappa} x^a \quad \text{and} \quad \{x^a, x^b\} = 0, \tag{4.80}$$

$$\{g_{ij}(p), g_{kl}(p)\} = 0, \tag{4.81}$$

and

$$\{x^\mu, g\} = -\tilde{X}^\mu g, \tag{4.82}$$

where $g_{ij}(p)$ are the entries of the matrix representing a pure AN(n) element in $T \times$ AN(n) and the last bracket can be written explicitly in terms of coordinates x^μ and the entries of g as $\{x^\mu, g_{ij}(p)\} = [\tilde{X}^\mu g]_{ij}(p)$, i.e., using the explicit parametrization of the matrix representation of g.[4] From (4.81), we can see that for any coordinates for the momentum group manifold, the Poisson brackets are

$$\{p_\mu, p_\nu\} = 0. \tag{4.84}$$

For illustrative purposes, we write down the explicit form of the Poisson brackets above for some specific parametrizations widely used in the literature. A pure $AN(n)$ group element g can be written as

$$g = e^{-\beta p_0 \tilde{X}^0} e^{-p_1 \tilde{X}^1 - p_2 \tilde{X}^2} e^{-(1-\beta) p_0 \tilde{X}^0}, \tag{4.85}$$

where the values for $0 \le \beta \le 1$ describe the different coordinates systems. Among them, as we will discuss in detail in the next chapter, $\beta = 0$ corresponds to the 'time-to-the-right' parametrization and in this case, $\{p_\mu\}$ are known as *bicrossproduct coordinates*, $\beta = 1$ corresponds to the time-to-the-left parametrization, and $\beta = \frac{1}{2}$ to the time-symmetric parametrization. The general group element (4.85) gives rise to the following general element $d = t\, g \in T \times AN(n)$

$$d = \begin{pmatrix} 1 & -\frac{1}{\kappa} e^{\frac{(1-\beta) p_0}{\kappa}} \mathbf{p}_n^{\mathsf{T}} \\ \mathbf{0}_n & e^{\frac{p_0}{\kappa}} \mathbb{1}_{n \times n} & \mathbf{x}_{(n+1)} \\ & \mathbf{0}_n^{\mathsf{T}} & 1 \end{pmatrix}, \tag{4.86}$$

where $\mathbf{p}_n = (p_1, \ldots, p_n)$ and $\mathbf{x}_{(n+1)} = (x^0, x^1, \ldots, x^n)$. Notice that the $AN(n)$ part is an upper-triangular matrix and this is a consequence of choosing the co-adjoint representation for the basis of its Lie algebra. From (4.82) and using the relation

---

[4]It is worth to mention that using the adjoint representation $\mathcal{R}$ for $\mathfrak{an}(n)$, we can describe the right decomposition of $d = g\, t \in T \times AN(n)$. The Poisson brackets are again obtained from (4.78) by just changing the sign of the r-matrix, $r \to -r$ and are given by (4.80), (4.81), and

$$\{x^\mu, g\} = g\, \tilde{X}^\mu. \tag{4.83}$$

$\{x^\mu, f(p)\} = \{x^\mu, p_\nu\}\frac{\partial f(p)}{\partial p_\nu}$, we find that the Poisson brackets for the different coordinates systems labelled by β are

$$\{x^0, p_0\} = 1,$$

$$\{x^0, p_a\} = -\frac{p_a}{\kappa}(1 - \beta),$$

$$\{x^a, p_0\} = 0,$$

$$\{x^a, p_b\} = \delta_b^a \, e^{\frac{p_0}{\kappa}\beta}. \tag{4.87}$$

It is worth to write down the relations for the cases $\beta = 0, \frac{1}{2}, 1$ and their first order expansions in $\frac{1}{\kappa}$. For the time-to-the-right case $\beta = 0$, the deformed brackets at *all orders* in κ read

$$\{x^0, p_0\} = 1, \quad \{x^0, p_a\} = -\frac{p_a}{\kappa}, \quad \{x^a, p_0\} = 0, \quad \{x^a, p_b\} = \delta_b^a, \tag{4.88}$$

reproducing the brackets (2.115) found in Chap. 2. The case $\beta = 1$ has the following Poisson brackets

$$\{x^0, p_0\} = 1, \quad \{x^0, p_a\} = 0, \quad \{x^a, p_0\} = 0, \quad \{x^a, p_b\} = \delta_b^a \, e^{\frac{p_0}{\kappa}}, \tag{4.89}$$

while for the time-symmetric case $\beta = \frac{1}{2}$, we have

$$\{x^0, p_0\} = 1, \quad \{x^0, p_a\} = -\frac{p_a}{2\kappa}, \quad \{x^a, p_0\} = 0, \quad \{x^a, p_b\} = \delta_b^a \, e^{\frac{p_0}{2\kappa}}. \tag{4.90}$$

Before closing the section, a remark is in order. We determined a symplectic Poisson structure on the phase space group manifold $\Gamma = T \times AN(n)$ using the Heisenberg double construction. The alternative Drinfeld double construction has the property of being compatible with the Lie group multiplication and indeed is related to the symmetries of the phase space [9]. Using (4.51) with the commutator, i.e., $\{d \overset{\otimes}{,} d\} = -[r, d \otimes d]_-$, we can find the Poisson brackets associated to the Drinfeld double structure. These are given again by (4.82) and (4.81) but now $\{x^\mu, g\} = 0$ for all μ, so the 'cross' brackets between position and momenta vanish identically.

4.4 Deforming Momentum Space to the Group SL(2, ℝ)

We now apply the same tools used for providing the group manifold $\Gamma = T \times AN(n)$ with a Poisson structure to the case, discussed at length in Chap. 2 in the context of gravity in $2 + 1$ dimensions, where momentum space is given by the Lie group $SL(2, \mathbb{R})$. In particular, we will show how the Cartesian product of the group of translations times the double cover of the three-dimensional Poincaré group $\mathbb{R}^{2,1} \times SL(2, \mathbb{R})$ can be turned into a proper phase space by using as the only input the Lie group structure of the (extended) momentum space $SL(2, \mathbb{R})$.

As in the previous section, we start from the Lie algebra $t \oplus \mathfrak{sl}(2, \mathbb{R})$ of the Lie group $\mathbb{R}^{2,1} \times SL(2, \mathbb{R})$ and the infinitesimal counterpart of the Poisson bivector, the co-commutator $\delta_{t \oplus \mathfrak{sl}(2,\mathbb{R})}$. Let us denote with $\{P_\mu\}$ and $\{\tilde{X}^\mu\}$ the generators of t and $\mathfrak{sl}(2, \mathbb{R})$, respectively, with Lie brackets

$$[P_\mu, P_\nu] = 0, \tag{4.91}$$

for all $\mu = 0, 1, 2$ and

$$[\tilde{X}^0, \tilde{X}^1] = -\frac{1}{\kappa}\tilde{X}^2, \quad [\tilde{X}^0, \tilde{X}^2] = \frac{1}{\kappa}\tilde{X}^1, \quad [\tilde{X}^1, \tilde{X}^2] = \frac{1}{\kappa}\tilde{X}^0, \tag{4.92}$$

which can be written in compact form

$$[\tilde{X}_\mu, \tilde{X}_\nu] = \frac{1}{\kappa}\varepsilon_{\mu\nu\sigma}\tilde{X}^\sigma, \tag{4.93}$$

with indices raised and lowered using the Minkowski metric and the totally skew-symmetric Levi-Civita pseudotensor such that $\varepsilon_{012} = 1$. The direct sum $t \oplus \mathfrak{sl}(2, \mathbb{R})$ can be turned into a Lie algebra by extending the inner product as in (4.55) and defining Lie brackets such that the product is ad-invariant. The resulting Lie algebra structure is given by the brackets (4.91, 4.92) together with

$$[P_0, \tilde{X}^1] = \frac{1}{\kappa}P_2, \quad [P_0, \tilde{X}^2] = -\frac{1}{\kappa}P_1, \quad [P_1, \tilde{X}^0] = \frac{1}{\kappa}P_2,$$
$$[P_1, \tilde{X}^2] = -\frac{1}{\kappa}P_0, \quad [P_2, \tilde{X}^0] = -\frac{1}{\kappa}P_1, \quad [P_2, \tilde{X}^1] = \frac{1}{\kappa}P_0. \tag{4.94}$$

We can now introduce co-commutators on $t \oplus \mathfrak{sl}(2, \mathbb{R})$ using an r-matrix analogous to (4.70). Denoting the complete set of generators $Z_A = \{P_\mu, \tilde{X}^\mu\}$, we have for the co-commutator the explicit relations

$$\delta_{t \oplus \mathfrak{sl}(2,\mathbb{R})}(\tilde{X}^\mu) = 0, \tag{4.95}$$

$$\delta_{t \oplus \mathfrak{sl}(2,\mathbb{R})}(P_0) = \frac{2}{\kappa}P_1 \wedge P_2, \quad \delta_{t \oplus \mathfrak{sl}(2,\mathbb{R})}(P_1) = \frac{2}{\kappa}P_0 \wedge P_2, \quad \delta_{t \oplus \mathfrak{sl}(2,\mathbb{R})}(P_2) = -\frac{2}{\kappa}P_0 \wedge P_1. \tag{4.96}$$

These co-commutators turn $t \oplus \mathfrak{sl}(2, \mathbb{R})$ into a Lie bi-algebra, i.e., a *classical double* which we denote as $\tilde{\mathscr{D}}(t)$. We can represent the generators as matrices using the co-adjoint representation, as in the previous section. In such representation, the generators Z_A can be written as 4×4-matrices as

$$P_\mu = \begin{pmatrix} 0_{3\times3} & \mathbf{u}_\mu \\ \mathbf{0}_3^T & 0 \end{pmatrix}, \tag{4.97}$$

where $\mathbf{u}_\mu$ is a three-component vector with 1 in the μth entry for $\mu = 0, \ldots, 2$ and

$$\tilde{X}^0 = \frac{1}{\kappa}\begin{pmatrix} 0 & 0 & 0 & 0 \\ 0 & 0 & 1 & 0 \\ 0 & -1 & 0 & 0 \\ 0 & 0 & 0 & 0 \end{pmatrix}, \quad \tilde{X}^1 = \frac{1}{\kappa}\begin{pmatrix} 0 & 0 & -1 & 0 \\ 0 & 0 & 0 & 0 \\ -1 & 0 & 0 & 0 \\ 0 & 0 & 0 & 0 \end{pmatrix}, \quad \tilde{X}^2 = \frac{1}{\kappa}\begin{pmatrix} 0 & 1 & 0 & 0 \\ 1 & 0 & 0 & 0 \\ 0 & 0 & 0 & 0 \\ 0 & 0 & 0 & 0 \end{pmatrix}.$$
(4.98)

A general element of the $T \times \mathsf{SL}(2, \mathbb{R})$ group manifold can be decomposed in terms of a pure translation and a pure $\mathsf{SL}(2, \mathbb{R})$ transformation as $d = tg$ with the following general matrix representation

$$d = \begin{pmatrix} \tilde{g} & \mathbf{x}_{n+1} \\ \mathbf{0}_{n+1}^{\mathsf{T}} & 1 \end{pmatrix},$$
(4.99)

where $\tilde{g} \in \mathsf{SL}(2, \mathbb{R})$ is a 3×3 matrix and

$$g = \begin{pmatrix} \tilde{g} & \mathbf{0}_{n+1} \\ \mathbf{0}_{n+1}^{\mathsf{T}} & 1 \end{pmatrix}, \quad t = \begin{pmatrix} \mathbb{1}_{(n+1)\times(n+1)} & \mathbf{x}_{n+1} \\ \mathbf{0}_{n+1}^{\mathsf{T}} & 1 \end{pmatrix},$$
(4.100)

are the matrices representing a pure Lorentz transformation and a pure translation, respectively.

As in the previous section, we can now use the Heisenberg double relation $\{d \overset{\otimes}{,} d\} = -[r, d \otimes d]_+$ to write down the Poisson brackets for the $T \times \mathsf{SL}(2, \mathbb{R})$ phase space. The general expressions for the Poisson bracket in terms of the coordinates x^μ appearing in (4.99) and of momenta are given by

$$\{x^0, x^1\} = -\frac{1}{\kappa}x^2, \quad \{x^0, x^2\} = \frac{1}{\kappa}x^1, \quad \text{and} \quad \{x^1, x^2\} = \frac{1}{\kappa}x^0, \quad (4.101)$$

$$\{g_{ij}(p), g_{kl}(p)\} = 0 \implies \{p_\mu, p_\nu\} = 0, \tag{4.102}$$

and

$$\{x^\mu, g\} = -\tilde{X}^\mu g. \tag{4.103}$$

Using the adjoint representation, instead of the co-adjoint one, the Poisson brackets are given again by (4.101) and (4.102) but now the mixed brackets read $\{x^\mu, g\} = -g\,\tilde{X}^\mu$.

It is instructive to focus on a given parametrization of the momentum group manifold. We consider the 'exponential coordinates' [10] for which $g \in \mathsf{SL}(2, \mathbb{R})$ is given by $g = e^{-p_\mu \tilde{X}^\mu}$, $\mu = 0, 1, 2$. The mass parameter $\kappa = 1/4\pi G$, as in Eq. 2.21 in Sect. 2.1, is determined by three-dimensional Newton's constant and in the limit

$G \to 0$, one recovers the usual flat momentum space $\mathbb{R}^{2,1}$. The matrix that describes the general group element $d \in T \times SL(2, \mathbb{R})$ is given by

$$
\begin{pmatrix}
-\frac{p_0^2-(p_1^2+p_2^2)\cos\frac{p}{\kappa}}{p^2} & -\frac{p_0p_1-p_0p_1\cos\frac{p}{\kappa}-p_2p\sin\frac{p}{\kappa}}{p^2} & -\frac{p_0p_2-p_0p_2\cos\frac{p}{\kappa}+p_1p\sin\frac{p}{\kappa}}{p^2} & x^0 \\[2mm]
\frac{p_0p_1-p_0p_1\cos\frac{p}{\kappa}+p_2p\sin\frac{p}{\kappa}}{p^2} & -\frac{p_1^2+(p_0^2-p_2^2)\cos\frac{p}{\kappa}}{p^2} & \frac{p_1p_2-p_1p_2\cos\frac{p}{\kappa}-p_0p\sin\frac{p}{\kappa}}{p^2} & x^1 \\[2mm]
\frac{p_0p_2-p_0p_2\cos\frac{p}{\kappa}-p_1p\sin\frac{p}{\kappa}}{p^2} & \frac{p_1p_2-p_1p_2\cos\frac{p}{\kappa}-p_0p\sin\frac{p}{\kappa}}{p^2} & -\frac{p_2^2+(p_0^2-p_1^2)\cos\frac{p}{\kappa}}{p^2} & x^2 \\[2mm]
0 & 0 & 0 & 1
\end{pmatrix},
$$
(4.104)

where $p^2 = -p_0^2 + p_1^2 + p_2^2$ and $p = \sqrt{|p^2|}$. The explicit, all order, relations for (4.103) in terms of the coordinates for the group (x^μ, p_μ) are rather involved and here we present these Poisson brackets at first order in the deformation parameter $\frac{1}{\kappa}$

$$\{x^0, p_0\} = 1, \qquad \{x^0, p_1\} = -\frac{1}{\kappa}\frac{p_2}{2}, \quad \{x^0, p_2\} = \frac{1}{\kappa}\frac{p_1}{2},$$

$$\{x^1, p_0\} = -\frac{1}{\kappa}\frac{p_2}{2}, \quad \{x^1, p_1\} = 1, \qquad \{x^1, p_2\} = -\frac{1}{\kappa}\frac{p_0}{2},$$

$$\{x^2, p_0\} = \frac{1}{\kappa}\frac{p_1}{2}, \qquad \{x^2, p_1\} = \frac{1}{\kappa}\frac{p_0}{2}, \qquad \{x^2, p_2\} = 1. \tag{4.105}$$

These relations can be written in a compact way as

$$\{x_\mu, p_\nu\} = \eta_{\mu\nu} + \frac{1}{\kappa}\,\varepsilon_{\mu\nu\alpha}\frac{p^\alpha}{2}. \tag{4.106}$$

We can also consider Cartesian coordinates on the group manifold by transforming the general group element in exponential coordinates (4.104) through the relations [10]

$$\tilde{p}_\mu = \frac{\sin\frac{p}{\kappa}}{p}p_\mu. \tag{4.107}$$

We find that the group element parametrized by Cartesian coordinate is represented by the following matrix

$$
d =
\begin{pmatrix}
\frac{\tilde{p}_0^2-(\tilde{p}_1^2+\tilde{p}_2^2)\sqrt{1-\frac{\tilde{p}^2}{\kappa^2}}}{\tilde{p}^2} & \frac{\tilde{p}_0\tilde{p}_1}{\left(1+\sqrt{1-\frac{\tilde{p}^2}{\kappa^2}}\right)\kappa^2} - \frac{\tilde{p}_2}{\kappa} & \frac{\tilde{p}_0\tilde{p}_2}{\left(1+\sqrt{1-\frac{\tilde{p}^2}{\kappa^2}}\right)\kappa^2} + \frac{\tilde{p}_1}{\kappa} & x^0 \\[4mm]
-\frac{\tilde{p}_0\tilde{p}_1}{\left(1+\sqrt{1-\frac{\tilde{p}^2}{\kappa^2}}\right)\kappa^2} - \frac{\tilde{p}_2}{\kappa} & \frac{\tilde{p}_1^2-(\tilde{p}_0^2-\tilde{p}_2^2)\sqrt{1-\frac{\tilde{p}^2}{\kappa^2}}}{\tilde{p}^2} & -\frac{\tilde{p}_1\tilde{p}_2}{\left(1+\sqrt{1-\frac{\tilde{p}^2}{\kappa^2}}\right)\kappa^2} - \frac{\tilde{p}_0}{\kappa} & x^1 \\[4mm]
-\frac{\tilde{p}_0\tilde{p}_2}{\left(1+\sqrt{1-\frac{\tilde{p}^2}{\kappa^2}}\right)\kappa^2} + \frac{\tilde{p}_1}{\kappa} & -\frac{\tilde{p}_1\tilde{p}_2}{\left(1+\sqrt{1-\frac{\tilde{p}^2}{\kappa^2}}\right)\kappa^2} + \frac{\tilde{p}_0}{\kappa} & \frac{\tilde{p}_2^2-(\tilde{p}_0^2-\tilde{p}_1^2)\sqrt{1-\frac{\tilde{p}^2}{\kappa^2}}}{\tilde{p}^2} & x^2 \\[4mm]
0 & 0 & 0 & 1
\end{pmatrix}
$$
(4.108)

where $\tilde{p}^2 = -\tilde{p}_0^2 + \tilde{p}_1^2 + \tilde{p}_2^2$. The Poisson brackets up to first order have the same form as in Equations (4.102), (4.101), and (4.105), that is, in compact form, we have $\{x_\mu, \tilde{p}_\nu\} = \eta_{\mu\nu} + \frac{1}{\kappa}\varepsilon_{\mu\nu\alpha}\frac{\tilde{p}^\alpha}{2}$. These Poisson brackets for the coordinates coincide with those found in Sect. 2.5 (Eq. 2.86). Indeed in that section, the Poisson brackets are given by $\{x^\mu, x^\nu\} = \frac{1}{\kappa}\varepsilon^{\mu\nu}{}_\rho x^\rho$, $\{p_\mu, p_\nu\} = 0$ with indices raised and lowered with a mostly plus Minkowski metric and $\varepsilon_{\mu\nu\alpha}$ is such that $\varepsilon_{012} = -1$.

4.5   Relativistic Hamiltonian Mechanics with Curved Momentum Space

In Part I, we argued that in 2+1 dimensions, gravity deforms particle kinematics by curving its momentum space in a particular way. In the case of the physical 3+1 spacetime dimensions, the analogous derivation is still missing (although, as we saw, there are some arguments that it may work as well), and therefore here, we will assume that the geometry of momentum space is arbitrary. This will make it possible to identify the geometric object that is necessary to make the transition from the standard formulation of the theory of relativistic particles to the one appropriate to the case of a non-trivial momentum space geometry.

To see how this generalization can be implemented, let us start with the discussion of the action of a free relativistic particle. This reads

$$S^0 = -\int_{-\infty}^{\infty} d\tau\, x^a \dot{p}_a + N\left(\eta^{ab} p_a p_b + m^2\right), \tag{4.109}$$

where the overdot denotes differentiation with respect to the parameter τ. The Lagrangian in (4.109) consists of two terms: the kinetic one $-x^a \dot{p}_a, a, b = 0, \ldots, 3$ and the mass-shell constraint $\eta^{ab} p_a p_b + m^2$ imposed by the Lagrange multiplier $N(\tau)$. It will be important for the later purposes to note that the term $\eta^{ab} p_a p_b$ is nothing but the square of the Minkowski distance between the point $\mathscr{P}$ in momentum space with coordinates p_a and the momentum space origin $\mathscr{O}$ with coordinates $p_a = 0$, calculated along the straight line joining these two points, i.e., along the geodesic of the Minkowski space geometry.

Let us note that the action (4.109) is manifestly invariant under global Lorentz transformations

$$\delta_\lambda x^a = \lambda^a{}_b x^b, \quad \delta_\lambda p_a = p_b \lambda^b{}_a, \quad \lambda^{ab} = -\lambda^{ba}, \tag{4.110}$$

and global translations

$$\delta_\xi x^a = \xi^a, \quad \delta_\xi p_a = \delta_\xi N = 0, \tag{4.111}$$

as well as under local τ reparametrization symmetry.

The equations of motion resulting from (4.109) are

$$\dot{p}_a = 0, \quad \dot{x}^a = 2N\eta^{ab} p_b, \quad \eta^{ab} p_a p_b = -m^2. \tag{4.112}$$

The first equation is the momentum conservation, the second relates velocity to momentum, while the third is the mass-shell condition.

The system of several non-interacting relativistic particles labelled by $\mathscr{I}$ is described by the action being a sum of the actions (4.109),

$$S^0_{free} = -\sum_{\mathsf{I}} \int d\tau \, x^a_{\mathsf{I}} \dot{p}^{\mathsf{I}}_a + N_{\mathsf{I}} \left(\eta^{ab} p^{\mathsf{I}}_a p^{\mathsf{I}}_b + m^2_{\mathsf{I}} \right) . \tag{4.113}$$

We can introduce particle interactions as follows. Let some number of worldliness meet at the interaction vertex, and let us assume that at the vertex, the momenta are conserved. One can further assume that for the worldlines corresponding to incoming particles, the parameter τ has the range from $-\infty$ to 0, while for the outgoing ones from 0 to ∞, and the interaction point corresponds to the $\tau = 0$ on each worldline, but from now on, we will not write the range explicitly.

To include the interaction, one adds to the action (4.113) an interaction term of the form [11]

$$S^0_{int} = z^a \, \mathscr{K}_a(p^{\mathsf{I}}) , \quad \mathscr{K}_a(p^{\mathsf{I}}) \equiv \widetilde{\sum_{\mathsf{I}}} p^{\mathsf{I}}_a, \tag{4.114}$$

where the tilde over sum indicates that the incoming momenta are taken with plus, while outgoing with minus signs. We assume that the total action is $S_{tot} = S_{free} + S_{int}$. Along the worldlines, the equations of motion following from this total action do not change and take the form (4.112) for each particle, and variation over z^a results in the momentum conservation rule at the vertex. However, since the worldlines are semi-infinite now, when varying the free action over momentum, from each particle action, we get a boundary term that gets combined with the variation of $\mathscr{K}_a$ leading to the condition

$$x^a_{\mathsf{I}}(0) = z^a \quad \forall \mathsf{I} . \tag{4.115}$$

This equation says that the 'local interaction coordinate' z^a is equal to the coordinates of the ends of the worldlines $x^a_{\mathsf{I}}(0)$. It is worth noticing that Eq. (4.115) is covariant under Lorentz transformations and invariant under translations, if we take $\delta_\xi z^a = \xi^a$. We see therefore that in special relativity, locality is *absolute*: if an event (interaction of particles) is local for one inertial observer, it is local for other inertial observers, in the sense that for all observers, $x^a_{\mathsf{I}}(0) = x^a_{\mathsf{J}}(0) = z^a$ for all I, J.

It should be noted that the absolute locality exhibited by the relativistic particles model relies on the 'correct' choice of the coordinates on the particles' phase space. Indeed, had we chosen another position coordinates $x^a \mapsto X^a \equiv M^a_b(p) \, x^b$, the theory would suffer from an apparent lack of locality: under translation, the worldlines would transform in a momentum-dependent way, i.e., instead of (4.111), we would have $\delta X^a = M^a_b(p) \, \xi^b$. We will return to this point later when discussing relative locality.

Let us now try to generalize the theory of free relativistic particle to the case of curved momentum space. Both kinetic term and the mass-shell condition of the action (4.109) will be affected by this generalization, so we will discuss these terms in turn.

To get an idea of how the kinetic term is to be modified, let us consider a case of the relativistic particle moving in curved spacetime. Usually, in the first-order formalism, the action is written as

$$S = \int d\tau \, p_\mu \dot{x}^\mu + N \left(g^{\mu\nu}(x) \, p_\mu \, p_\nu + m^2 \right) .$$

Starting from this action, one can easily check that the resulting equation of motion is the equation of geodesics of the metric $g_{\mu\nu}(x)$.

Notice that in this formulation, the momentum p_μ has a 'curved' spacetime index, even if the momentum space is flat, and that, as a result, the mass-shell relation becomes spacetime point dependent, which is an undesirable feature of this formulation. There is however a different form of this same action, which manifestly exhibits the fact that the momentum space is flat, while the spacetime is curved. To get this form, one replaces the momentum labelled with the curved index with the 'flat index' one, i.e., $p_\mu \rightarrow p_a \equiv e_a^\mu(x) \, p_\mu$, where $e_a^\mu(x)$ satisfying $\eta^{ab} e_a^\mu(x) e_b^\nu(x) = g^{\mu\nu}(x)$ is the (inverse) tetrad associated with the metric $g_{\mu\nu}(x)$. In terms of these new variables, the action of a relativistic particle moving on curved background takes the form

$$S = \int d\tau \, p_a \, e_\mu^a(x) \, \dot{x}^\mu + N \left(\eta^{ab} \, p_a \, p_b + m^2 \right) . \tag{4.116}$$

Now it is clear how to proceed in an 'upside-down' case, when it is the momentum space that becomes curved, while the spacetime remains flat[5]: in the kinetic term, one has to label the momentum with curved momentum space index $\alpha, \beta, \ldots$ and to use the momentum space tetrad, so that

$$L_{kin} = -x^a \, E_a^\alpha(p) \, \dot{p}_\alpha . \tag{4.117}$$

It should be stressed that in order to be able to construct the particle Lagrangian with a non-trivial momentum space tetrad, one must have in disposal a physical scale of momentum κ, because, by definition, the momentum space tetrad is a dimensionless function of momentum, so that it can depend on momenta only through the dimensionless combination p/κ. In the case of the theories that we described in Part I, such scale has been provided by the theory of gravity, but it is possible just to assume that the non-trivial geometry of momentum space is the fundamental feature of the theory, in which case the presence of the scale follows from the first principles.

In order to complete the construction of the action of a particle with curved momentum space, we must append the kinetic Lagrangian (4.117) with an appropriate mass-shell condition. The form of this condition might be dictated by construction,

---

[5]The generalization to the case of both momentum space and spacetime being curved is not-trivial. The reader can find a detailed description of this construction in the paper [12].

as it was in the cases discussed in Part I, but if such information is not available, in order to preserve the symmetries of the kinetic term (see below), one can proceed as follows. Take the momentum space metric constructed from the tetrad $G^{\alpha\beta}(p) \equiv \eta_{ab} E_a^\alpha(p) E_b^\beta(p)$. Then we postulate that the mass-shell condition takes the form

$$\mathscr{C}(p) + m^2 = 0,$$

where $\mathscr{C}(p)$ is the square of the distance between the origin of momentum space $\mathscr{O} : p_\alpha = 0$ and the point $\mathscr{P}$ with coordinates p_α, calculated along the geodesics of the metric $G^{\alpha\beta}$ joining these two points. This construction clearly reproduces the standard form of the mass-shell condition in the case of the flat momentum space. Notice also that instead of using the geodesic distance, one could use some simpler expression for $\mathscr{C}$, provided that the solution of the mass-shell condition is the same (up to the redefinition of the mass parameter m, which can always be done when the mass scale is present.)

In this way, we construct the action of a single particle with curved momentum space, which reads

$$S = -\int d\tau \, x^a \, E_a^\alpha(p) \, \dot{p}_\alpha - N \left(\mathscr{C}(p) + m^2 \right). \qquad (4.118)$$

The equations of motion following from this action are

$$\dot{p}_\alpha = 0,$$
$$\dot{x}^a \, E_a^\alpha + N \frac{\partial \mathscr{C}}{\partial p_\alpha} = 0, \qquad (4.119)$$
$$\mathscr{C}(p) + m^2 = 0,$$

where in deriving the first equation, we used the fact that the momentum space tetrad is invertible and we used the first equation in the second. We see that momenta are conserved, as in the case of flat momentum space, but the relation between four velocities and momenta is being modified. In fact, it is usually hard to explicitly invert the momentum—velocity relation, which could be used to obtain the second-order form of the action (4.118).

Let us now compute the Poisson brackets of the theory defined by (4.118). To this end, it is convenient to introduce the positions in the cotangent space of the momentum space (depending on p)

$$x^\alpha(p) = x^a \, E_a^\alpha(p)$$

in terms of which the Poisson brackets are classical

$$\{x^\alpha, p_\beta\} = \delta_\beta^\alpha, \quad \{x^\alpha, x^\beta\} = \{p_\alpha, p_\beta\} = 0.$$

Then for the physical position variables, we find

$$\left\{ x^a, p_\beta \right\} = E^a_\beta(p) \tag{4.120}$$

$$\left\{ x^a, x^b \right\} = E^a_\alpha E^b_\beta \left(E^{\alpha,\beta}_c - E^{\beta,\alpha}_c \right) x^c, \tag{4.121}$$

with momenta having vanishing Poisson bracket. We see that the Poisson bracket of two positions does not vanish if the momentum space tetrad is non-trivial, and therefore, the system, after quantization, will correspond to a non-commutative spacetime.

Let us now investigate the symmetries of the deformed particle action (4.118). First of all, this action is invariant under deformed, momentum -ependent translations

$$\delta_\xi x^a = E^a_\alpha(p)\, \xi^\alpha, \quad \delta_\xi p_\alpha = \delta_\xi N = 0. \tag{4.122}$$

One can check that the action of two infinitesimal translations commutes. This can be seen as follows. Consider the commutator

$$[\delta_{\xi_1}, \delta_{\xi_2}]\, x^a = \delta_{\xi_1} \left(\delta_{\xi_2}\, x^a \right) - 1 \leftrightarrow 2 = 0,$$

because p does not transform under translations.

We can now calculate the conserved Noether charges associated with translations. To this end, we use the general construction which goes as follows. Assume that the parameter of infinitesimal translation is τ-dependent $\xi = \xi(\tau)$. Then the conserved Noether charge is the coefficient proportional to the time derivative $\dot\xi$ in the variation of the action (when total time derivative is disregarded). To see this, consider the action depending on some variables $\phi^{(a)}(\tau)$ (positions and momenta)

$$S = \int d\tau\, L(\phi, \dot\phi),$$

invariant under infinitesimal global symmetry parametrized by (τ-independent) infinitesimal parameters ξ^α, i.e.,

$$\delta\phi^{(a)} = \xi^\alpha\, R^{(a)}_\alpha(\phi, \dot\phi).$$

The action is invariant if the variation of the Lagrangian is a total τ derivative

$$\delta L = \frac{\partial L}{\partial \phi^{(a)}} \xi^\alpha R^{(a)}_\alpha + \frac{\partial L}{\partial \dot\phi^{(a)}} \xi^\alpha \dot{R}^{(a)}_\alpha = \xi^\alpha \dot{\Lambda}_\alpha. \tag{4.123}$$

This equality must hold even if the equations of motion are not satisfied; if however we assume that the equations of motion hold, rewriting the second term as a total time derivative minus the time derivative of $\partial L/\partial \dot\phi^{(a)}$ and using equations of motion, we immediately find that there is a conserved quantity, the Noether charge, associated with the symmetry

$$\frac{d}{d\tau} \left(\frac{\partial L}{\partial \dot\phi^{(a)}} R^{(a)}_\alpha - \Lambda \right) = 0. \tag{4.124}$$

For the action (4.118) and translational symmetry, we find that the functions R associated with momenta are zero, while $\Lambda = p$, so that the components of momenta p_α are the sought conserved Noether charges, as expected, in agreement with the equations of motion (4.119).

The important property of the symmetry (4.122) is that worldlines of particles with different momenta are translated differently, so that, naively, one would expect that the worldlines meeting at a point for one observer do not meet for another. In other words, locality of events (which can be only physically defined as the crossing point of some particles worldlines[6]) ceases to be absolute and becomes relative.

The Lorentz invariance of the action (4.118) is not that easy to see, so we present the construction step by step, discussing all details.

The Lorentz symmetry algebra consists of two sets of generators, rotations M_i and boosts N_i, which satisfy

$$[M_a, M_b] = \varepsilon_{ab}{}^c M_c \,, \quad [M_a, N_b] = \varepsilon_{ab}{}^c N_c \,, \quad [N_a, N_b] = -\varepsilon_{ab}{}^c M_c \,. \quad (4.125)$$

The corresponding infinitesimal symmetry transformations δ_ρ and δ_λ should satisfy the same algebra.

Let us investigate first the symmetry properties of the term $E_a^\alpha(p)\,\dot{p}_\alpha$ in the action (4.118). Since we want the mass-shell condition to be invariant, and it is constructed as a geodesic distance between points $G^{\alpha\beta}(p)$, we demand that

$$\delta_{\rho/\lambda}\left(G^{\alpha\beta}\,\dot{p}_\alpha\,\dot{p}_\beta\right) = \delta_{\rho/\lambda}\left(\eta^{ab}\,E_a^\alpha\,E_b^\beta\,\dot{p}_\alpha\,\dot{p}_\beta\right) = 2\eta^{ab}\,E_a^\alpha\,\dot{p}_\alpha\,\delta_{\rho/\lambda}\left(E_b^\beta\,\dot{p}_\beta\right) = 0\,. \quad (4.126)$$

It is usually assumed that the action of the rotational subgroup of the Lorentz group is purely classical, without any deformations. The reason is that the rotational invariance is pretty well checked experimentally and so far no deviation from the standard picture has been observed. Notice that the situation with boosts is completely different and we checked symmetry of physical systems under boosts only for a very moderate range of small boosts parameters.

Following this discussion, we assume that rotations act classically, i.e.,

$$\delta_\rho E_0^\alpha(p)\,\dot{p}_\alpha = 0\,, \quad \delta_\rho E_i^\alpha(p)\,\dot{p}_\alpha = \varepsilon_{ij}{}^k \rho^j\,E_k^\alpha(p)\,\dot{p}_\alpha\,. \quad (4.127)$$

Clearly, the condition (4.126) is satisfied and it is sufficient to append these transformations with

$$\delta_\rho x^0 = 0\,, \quad \delta_\rho x^i(p) = \varepsilon^i{}_{jk}\rho^j\,x^k \quad (4.128)$$

to ensure invariance of the action. Notice that

$$[\delta_{\rho_1}, \delta_{\rho_2}]E_i^\alpha(p)\,\dot{p}_\alpha = \delta_{\rho_1}\left(\varepsilon_{ij}{}^k \rho_2^j\,E_k^\alpha(p)\,\dot{p}_\alpha\right) - 1 \leftrightarrow 2 = \varepsilon_{ij}{}^k\,\rho^j\,E_k^\alpha(p)\,\dot{p}_\alpha\,,$$

---

[6]For example, the statement I see the table means really that the worldline of the table crossed the worldlines of photons that hit my retina.

with

$$\rho^j = \varepsilon^j{}_{kl}\, \rho_2^k \rho_1^l .$$

Therefore, the infinitesimal rotations δ_ρ form indeed the representation of the rotational group (cf. the first commutator in (4.125).)

Let us now turn to the boosts. As discussed above, these transformations might become deformed, and we have to find the most general boost transformations δ_λ which are representations of the Lorentz algebra (4.125) and satisfy the condition (4.126). Then we can always adjust the transformations of positions so as to make the action invariant.

It could be checked that the most general boost transformations that leaves the mass-shell condition invariant has the form

$$\delta_\lambda E_0^\alpha(p)\, \dot{p}_\alpha = \lambda^i\, f(p)\, E_i^\alpha(p)\, \dot{p}_\alpha , \tag{4.129}$$

$$\delta_\lambda E_i^\alpha(p)\, \dot{p}_\alpha = \lambda_i\, f(p)\, E_0^\alpha(p)\, \dot{p}_\alpha + \varepsilon_{ij}{}^k\, \lambda^l\, g_l^j(p)\, E_k^\alpha(p)\, \dot{p}_\alpha . \tag{4.130}$$

In this formula, $f(p)$ and $g^j(p)$ are arbitrary momentum-dependent functions, such that the boost transformation is a representation of the second and the third commutational relation in (4.125). Unfortunately, not much could be deduced from Eqs. (4.129) and (4.130) without having an explicit form of the momentum space tetrads.

It follows that x^a transform under momentum-dependent Lorentz transformations as components of a Lorentz vector as well

$$\delta_\lambda x^0 = -x^i \bar{\lambda}_i(p) , \quad \delta_\lambda x^i = -x^0 \bar{\lambda}^i(p) + \varepsilon^i{}_{jk}\, \bar{\rho}^j(p)\, x^k , \tag{4.131}$$

and the action (4.118) is Lorentz invariant.

This concludes our discussion of the general theory of free particles with curved momentum space.

4.6 κ-Poincaré Particle—The Free Case

Till now we discussed the particles' model in a rather abstract way, without referring to any particular example. Let us therefore turn now to a specific model, which has its roots in κ-Poincaré algebra and κ-Minkowski space constructions [13–16]. In what follows, we will borrow from the presentations in Refs. [17,18].

Let us start with the momentum space of a κ-particle, which, as we saw in the previous section, is given by the four-dimensional Lie group AN(3), whose Lie algebra generators satisfy

$$[X^0, X^i] = \frac{i}{\kappa}\, X^i . \tag{4.132}$$

Sometimes X^a above are interpreted as positions in a non-commutative spacetime; such spacetime is known under the name[7] of 'κ-Minkowski space' [15]. For our purposes, the relevant matrix representation of this Lie algebra is the five-dimensional one, in which case we have

$$X^0 = -\frac{i}{\kappa}\begin{pmatrix} 0 & 0 & 1 \\ 0 & 0 & 0 \\ 1 & 0 & 0 \end{pmatrix} \quad \mathbf{X} = \frac{i}{\kappa}\begin{pmatrix} 0 & \boldsymbol{\varepsilon}^T & 0 \\ \boldsymbol{\varepsilon} & 0 & \boldsymbol{\varepsilon} \\ 0 & -\boldsymbol{\varepsilon}^T & 0 \end{pmatrix}, \tag{4.133}$$

where bold fonts are used to denote space components of a four-vector and $\boldsymbol{\varepsilon}$ is a three-dimensional vector with a single unit entry, e.g., $\varepsilon^1 = (1,0,0)$. Let us now consider a group element of AN(3) (which can be seen as an 'ordered plane wave on κ-Minkowski space'[20]), which, since it represents a group-valued momentum, we will denote by Π, as we did in Part I

$$\Pi(p) = e^{ip_i X^i} e^{ip_0 X^0}. \tag{4.134}$$

In the representation (4.133), this is a 5×5 matrix which acts on five-dimensional Minkowski space as a linear transformation. One finds

$$\exp(ip_0 X^0) = \begin{pmatrix} \cosh\frac{p_0}{\kappa} & \mathbf{0} & \sinh\frac{p_0}{\kappa} \\ \mathbf{0} & \mathbf{1} & \mathbf{0} \\ \sinh\frac{p_0}{\kappa} & \mathbf{0} & \cosh\frac{p_0}{\kappa} \end{pmatrix}, \quad \exp(ip_i X^i) = \begin{pmatrix} 1+\frac{\mathbf{p}^2}{2\kappa^2} & \frac{\mathbf{p}}{\kappa} & \frac{\mathbf{p}^2}{2\kappa^2} \\ \frac{\mathbf{p}}{\kappa} & 1 & \frac{\mathbf{p}}{\kappa} \\ -\frac{\mathbf{p}^2}{2\kappa^2} & -\frac{\mathbf{p}}{\kappa} & 1-\frac{\mathbf{p}^2}{2\kappa^2} \end{pmatrix},$$

where $\mathbf{1}$ is the unit 3×3 matrix from which we find

$$\Pi(p) = \begin{pmatrix} \frac{\bar{P}_4}{\kappa} & e^{-p_0/\kappa}\frac{\mathbf{P}}{\kappa} & \frac{P_0}{\kappa} \\ \frac{\mathbf{P}}{\kappa} & 1 & \frac{\mathbf{P}}{\kappa} \\ \frac{\bar{P}_0}{\kappa} & -e^{-p_0/\kappa}\frac{\mathbf{P}}{\kappa} & \frac{P_4}{\kappa} \end{pmatrix}. \tag{4.135}$$

To construct the group manifold of the group AN(3), we now take a special point in Minkowski space, which will become the momentum space origin $\mathcal{O}$ with coordinates $(0,\ldots,0,\kappa)$ and act on it with the matrix Π (4.135). As a result, we get a point in

---

[7] In the κ-Poincaré literature, this algebra is usually called the κ-Minkowski algebra and is defined as a dual (in the Hopf algebraic sense) to the κ-Poincaré Hopf algebra, see Refs. [15,16,19] for more details and further references.

the five-dimensional Minkowski space, being in the one to one correspondence with the group element Π, whose coordinates (P_0, P_i, P_4) are[8]

$$P_0(p_0, \mathbf{p}) = \kappa \sinh \frac{p_0}{\kappa} + \frac{\mathbf{p}^2}{2\kappa} e^{p_0/\kappa},$$

$$P_i(p_0, \mathbf{p}) = p_i e^{p_0/\kappa}, \tag{4.136}$$

$$P_4(p_0, \mathbf{p}) = \kappa \cosh \frac{p_0}{\kappa} - \frac{\mathbf{p}^2}{2\kappa} e^{p_0/\kappa}.$$

It is easy to check that

$$- P_0^2 + \mathbf{P}^2 + P_4^2 = \kappa^2, \tag{4.137}$$

and therefore, the group AN(3) is isomorphic, as a manifold to four-dimensional de Sitter space. Actually, this group is not the whole of de Sitter space, but rather a half of it because it follows from (4.136) that

$$P_0 + P_4 = e^{p_0/\kappa} > 0, \quad P_4 \equiv \sqrt{\kappa^2 + P_0^2 - \mathbf{P}^2} > 0. \tag{4.138}$$

To construct the free action with the group AN(3) as a momentum space, we can proceed as in Part I, or, equivalently we recall that in this case, there exists a canonical form of the kinetic term provided by the Kirillov symplectic form [21]. It can be constructed as follows. Since the positions belong to the cotangent space of the momentum space, they can be naturally associated with elements of a dual to the Lie algebra, which is a linear space spanned by the basis σ_a with the pairing

$$\langle \sigma_a, X^b \rangle = \frac{1}{i} \delta_a^b. \tag{4.139}$$

Now having a group element $\Pi(p)$ (4.134), we know that $\Pi^{-1} \dot{\Pi}$ and $\dot{\Pi} \Pi^{-1}$ belong to the Lie algebra AN(3) and the kinetic term of the action can be defined by the pairing of one of these elements with the position variables $x^a \sigma_a$, belonging to the dual of the AN(3) algebra. Using (4.134) with the help of Baker–Campbell–Hausdorff formula, we find

$$\Pi^{-1} \dot{\Pi} = i \dot{p}_0 X^0 + i \dot{p}_i e^{-i p_0 X^0} X^i e^{i p_0 X^0} = i \dot{p}_0 X^0 + e^{p_0/\kappa} i \dot{p}_i X^i, \tag{4.140}$$

$$\dot{\Pi} \Pi^{-1} = i \dot{p}_i X^i + i \dot{p}_0 e^{i p_i X^i} X^0 e^{-i p_i X^i} = i \left(\dot{p}_i - \frac{p_i}{\kappa} \dot{p}_0 \right) X^i + i \dot{p}_0 X^0. \tag{4.141}$$

---

[8]If we choose another point as $\mathcal{O}$, we would get different realizations of the group $AN(3)$. For example, if $\mathcal{O} = (\kappa, \ldots, 0)$, we would get the Euclidean realization and for $\mathcal{O} = (\kappa, 0, 0, 0, \kappa)$, the light-cone one. See [7] for more details of this construction.

Thus, the kinetic term of the action is

$$S^{kin} = -\int d\tau \, x^a \langle \sigma_a, \Pi^{-1} \dot{\Pi} \rangle = -\int d\tau \, x^0 \dot{p}_0 + e^{p_0/\kappa} \, x^i \, \dot{p}_i \,, \qquad (4.142)$$

in the first case (4.140), and

$$S^{kin} = -\int d\tau \, x^a \langle \sigma_a, \dot{\Pi} \Pi^{-1} \rangle = -\int d\tau \, x^0 \dot{p}_0 + x^i \, \dot{p}_i - \frac{1}{\kappa} x^i \, p_i \, \dot{p}_0 \,, \quad (4.143)$$

in the case (4.141). Notice that the action (4.143) can be obtained from (4.142) if we change the definition of spacial momenta $p_i \to e^{-p_0/\kappa} \, p_i$ and therefore in what follows, we will analyse only the first. The reader can easily reproduce all the steps below in the case of the second action or just find the corresponding formulas by changing the momentum variables.

Recalling (4.118) from (4.142), we can immediately read off the components of the momentum space tetrad

$$E^0_0 = 1 \,, \quad E^j_i = e^{p_0/\kappa} \, \delta^j_i \,, \qquad (4.144)$$

and the line element takes the form

$$ds^2 = -dp_0^2 + e^{2p_0/\kappa} \, d\mathbf{p}^2 \,, \qquad (4.145)$$

where $d\mathbf{p}^2 \equiv dp_i dp_i$, or

$$G^{00} = -1 \,, \quad G^{ii} = e^{2p_0/\kappa} \,, \qquad (4.146)$$

which is nothing but the metric of de Sitter space in 'flat cosmological' coordinates. It should be noticed that the metric (4.146) can be obtained as an induced metric from the five-dimensional Minkowski line element $ds^2 = -dP_0^2 + d\mathbf{P}^2 + dP_4^2$ when (4.136) is used.

Having the metric, we can calculate the distance function, which is used to define the mass-shell condition; it reads [18,22]

$$\mathscr{C}(p) = \kappa^2 \operatorname{arccosh} \frac{P_4}{\kappa}$$

so that the mass-shell condition reads

$$\cosh \frac{p_0}{\kappa} - \frac{\mathbf{p}^2}{2\kappa^2} e^{p_0/\kappa} = \cosh \frac{m}{\kappa} \,. \qquad (4.147)$$

Using (4.142) and (4.147), we construct the free particle action

$$S^{\kappa P}_{free} = -\int d\tau \left(x^0 \dot{p}_0 + e^{p_0/\kappa} \, x^i \, \dot{p}_i + 2N \left[\kappa^2 \cosh \frac{p_0}{\kappa} - \frac{\mathbf{p}^2}{2} e^{p_0/\kappa} - \kappa^2 \cosh \frac{m}{\kappa} \right] \right) \,, \qquad (4.148)$$

where we replaced N with $2N$ so that the action $S_{free}^{\kappa P}$ reduces to the standard relativistic particle action in the limit $\kappa \to \infty$. It is worth noticing that the action (4.148) leads to a non-trivial Poisson brackets algebra, which, of course, is a special case of the general formulas (4.120) and (4.121)

$$\left\{x^0, x^i\right\} = -\frac{1}{\kappa} x^i, \quad \left\{x^i, x^j\right\} = 0,$$

$$\left\{x^0, p_0\right\} = 1, \quad \left\{x^i, p_j\right\} = e^{-p_0/\kappa} \delta_j^i, \quad \left\{x^0, p_i\right\} = \left\{x^i, p_0\right\} = 0, \quad (4.149)$$

with p_α having vanishing Poisson brackets.

The equations of motion following from (4.148) have the form

$$\dot{p}_\alpha = 0,$$

$$\dot{x}^i = -2N\, p_i,$$

$$\dot{x}^0 = 2N\left(\kappa \sinh \frac{p_0}{\kappa} - \frac{\mathbf{p}^2}{2\kappa} e^{p_0/\kappa}\right), \quad (4.150)$$

supplemented by the mass-shell condition (4.147), where in the equations for x^a, we omitted terms proportional to τ derivatives of momenta. It is worth noticing that the velocity of massless particles

$$\mathbf{v}^2 = \left|\frac{\dot{x}^i}{\dot{x}^0}\right|^2 = 1$$

so that the velocity of light is independent of the energy of photons [23].

Let us now turn to the symmetries of the action (4.148). It follows from the general theory presented above that the κ-Poincaré particle possesses, like an ordinary relativistic particle, a ten-dimensional group of relativistic symmetries; however, now the symmetries are *deformed*. As we will see, there is also another potential problem with Lorentz symmetry because the orbits of the naive action of Lorentz group do not always belong to the momentum space, so the action is not transitive. This unpleasant property of the naive Lorentz group action will force us to invent a more ingenious definition of it, which will make use of the notion of the antipode and will be our first encounter with the Hopf algebra terminology.

Let us start with translations. The explicit form of these transformations that leave the action (4.148) invariant can be easily deduced from the general discussion presented above, see (4.122), and reads

$$\delta x^0 = \xi^0, \quad \delta x^i = \xi^i\, e^{-p_0/\kappa}. \quad (4.151)$$

It follows that the conserved Noether charges associated with the translational symmetry are the components of the momentum, as usual.

Let us now turn to Lorentz symmetry. There are two ways to derive it. First one can use the general formulas derived in the previous section (4.129), (4.130), which in the case of the tetrad (4.144) take the form

$$\delta_\lambda \, \dot{p}_0 = \lambda^i \, f(p) \, e^{p_0/\kappa} \, \dot{p}_i \,, \tag{4.152}$$

$$\delta_\lambda \left(e^{p_0/\kappa} \, \dot{p}_i \right) = \lambda_i \, f(p) \, \dot{p}_0 + \varepsilon_{ij}{}^k \, \lambda^l \, g_l^j(p) \, e^{p_0/\kappa} \, \dot{p}_k \,. \tag{4.153}$$

Another way is to use five-dimensional Minkowski coordinates (4.136), of which P_0, P_i transform under boosts as components of a Lorentz (four-) vector, while P_4 remains invariant, to wit

$$\delta_\lambda \, P_0 = \lambda^i \, P_i \,, \quad \delta_\lambda \, P_i = \lambda_i \, P_0 \,, \quad \delta_\lambda \, P_4 = 0 \,, \tag{4.154}$$

and derive from these equations the formulas for $\delta_\lambda \, p_0$ and $\delta_\lambda \, p_i$. This procedure is simpler and we will follow it here.

Taking variation of the first and last equation in (4.136), using (4.154), and then adding the resulting equations, we find that

$$\delta_\lambda \, p_0 = \lambda^i \, p_i \,. \tag{4.155}$$

Then from the middle equation in (4.136), we get

$$\delta_\lambda \, p_i = \lambda_i \left(\frac{\kappa}{2} \left(1 - e^{-2p_0/\kappa} \right) + \frac{\mathbf{p}^2}{2\kappa} \right) - \frac{1}{\kappa} \, p_i \, \lambda^j \, p_j \,. \tag{4.156}$$

We leave it to the reader to check that the transformations (4.155) and (4.156) satisfy Eqs. (4.152) and (4.153) with an appropriate choice of $f(p)$ and $g_l^j(p)$.

The transformations (4.155) and (4.156) leave the mass-shell constraint invariant by construction. We still have to accompany them with the transformations rules for positions, so as to make the action (4.148) invariant. The variation of the Lagrangian in (4.148) is

$$\delta_\lambda x^0 \dot{p}_0 + x^0 \delta_\lambda \dot{p}_0 + \delta_\lambda x^i \, e^{p_0/\kappa} \, \dot{p}_i + x^i \delta_\lambda \left(e^{p_0/\kappa} \, \dot{p}_i \right) = 0 \,.$$

Since $\dot{p}_0$ and $\dot{p}_i$ are independent, the expressions multiplying them must vanish independently. Collecting first the terms with $\dot{p}_0$ factor, we get

$$\delta x^0 = -\lambda_i x^i \, e^{-p_0/\kappa} \,. \tag{4.157}$$

Then vanishing of the terms with the $\dot{p}_i$ factor leads to

$$\delta x^i = -\lambda^i x^0 \, e^{-p_0/\kappa} - \frac{1}{\kappa} \left(x^j \lambda_j \, p^i - x^j p_j \, \lambda^i \right) \,. \tag{4.158}$$

This completes our description of the properties of the free single κ-deformed particle. Let us now turn to the description of a system of two κ-deformed particles.

4.7   The System of Two κ-Deformed Particles

In this section, we will use the lesson learned in the construction of the deformed multi-particle Lagrangian in the case of particles 2+1 gravity to define the κ-deformed two particles' action and to investigate its properties. The presentation here will be based on [24] and the reader could consult this paper for more technical details.

In accordance with the results of Sect. 2.5, we define the Lagrangian of two κ-deformed particles to be (see (2.90))

$$\mathscr{L}^{kin}_{1\oplus2} = \left\langle \dot{\Pi}_{(1)} \Pi^{-1}_{(1)}, x_{(1)} \right\rangle + \left\langle \dot{\Pi}_{(2)} \Pi^{-1}_{(2)}, x_{(2)} \right\rangle + \left\langle \Pi_{(1)} \dot{\Pi}_{(2)} \Pi^{-1}_{(2)} \Pi^{-1}_{(1)} - \dot{\Pi}_{(2)} \Pi^{-1}_{(2)}, x_{(1)} \right\rangle,$$
$$(4.159)$$

where the lower index labels the particle. Using the expression for each group-valued momentum

$$\Pi(p) = e^{ip_i X^i} e^{ip_0 X^0} \tag{4.160}$$

and the pairing (4.139), we find the kinetic part of the Lagrangian of two κ-deformed particles

$$\mathscr{L}^{kin}_{1\oplus2} = \left(x^0_{(1)} + \frac{1}{\kappa} x^j_{(1)} p_{(1)j} \right) \dot{p}_{(1)0} + x^j_{(1)} \dot{p}_{(1)j}$$
$$+ \left(x^0_{(2)} + \frac{1}{\kappa} x^j_{(2)} p_{(2)j} - \frac{1}{\kappa} x^j_{(1)} \left(\left(1 - e^{-\frac{1}{\kappa} p_{(1)0}} \right) p_{(2)j} - p_{(1)j} \right) \right) \dot{p}_{(2)0}$$
$$+ \left(x^j_{(2)} - x^j_{(1)} \left(1 - e^{-\frac{1}{\kappa} p_{(1)0}} \right) \right) \dot{p}_{(2)j}. \tag{4.161}$$

To complete the construction of the Lagrangian for the system of two κ-deformed particles, we have to impose the mass-shell constraints on each particle. The mass-shell relation is given by κ-Poincaré mass Casimir

$$\mathscr{C}_{(a)} = 4\kappa^2 \sinh \left(\frac{p_{(a)0}}{2\kappa} \right)^2 - e^{p_{(a)0}/\kappa} \mathbf{p}^2_{(a)}. \tag{4.162}$$

Thus, the final two-particle Lagrangian is

$$\mathscr{L}_{1\oplus2} = \mathscr{L}^{kin}_{1\oplus2} + \lambda_{(1)} \left(\mathscr{C}_{(1)} - m^2_{(1)} \right) + \lambda_{(2)} \left(\mathscr{C}_{(2)} - m^2_{(2)} \right). \tag{4.163}$$

From the Lagrangian (4.163), we derive equations of motion following from the variation of positions $x^\mu_{(a)}$

$$\dot{p}_{(1)\mu} = \dot{p}_{(2)\mu} = 0, \tag{4.164}$$

which are the usual momentum conservation conditions. Varying over momenta, we get

$$\dot{x}^0_{(1)} = 2\lambda_{(1)} \left(\kappa \sinh \left(\frac{P_{(1)0}}{\kappa} \right) + \frac{1}{2\kappa} e^{P_{(1)0}/\kappa} \mathbf{p}^2_{(1)} \right), \qquad \dot{x}^j_{(1)} = -2\lambda_{(1)} e^{P_{(1)0}/\kappa} p_{(1)j},$$

$$\dot{x}^0_{(2)} = 2\lambda_{(2)} \left(\kappa \sinh \left(\frac{P_{(2)0}}{\kappa} \right) + \frac{1}{2\kappa} e^{P_{(2)0}/\kappa} \mathbf{p}^2_{(2)} \right) + 2\lambda_{(1)} \frac{1}{\kappa} e^{P_{(1)0}/\kappa} \mathbf{p}^2_{(1)},$$

$$\dot{x}^j_{(2)} = -2\lambda_{(2)} e^{P_{(2)0}/\kappa} p_{(2)j} - 2\lambda_{(1)} \left(e^{P_{(1)0}/\kappa} - 1 \right) p_{(1)j},$$

$$\tag{4.165}$$

which provide us with (deformed) relations between velocities and momenta. This relations possess an interesting property, reflecting the non-trivial structure of the coupling between particles resulting from the deformation. Namely the velocity of the first particle depends only on the first momentum, while the velocity of the second depends on the momenta of both particles.

Starting from the Lagrangian (4.163), we can compute the symplectic structure and the associated Poisson brackets, which read

$$\left\{ P_{(a)0}, x^0_{(a)} \right\} = 1, \quad \left\{ P_{(a)0}, x^j_{(a)} \right\} = 0, \tag{4.166}$$

$$\left\{ P_{(a)j}, x^0_{(a)} \right\} = -\frac{1}{\kappa} P_{(a)j}, \quad \left\{ P_{(a)j}, x^k_{(a)} \right\} = \delta^k_j, \tag{4.167}$$

$$\left\{ P_{(1)0}, x^\nu_{(2)} \right\} = 0, \quad \left\{ P_{(1)j}, x^0_{(2)} \right\} = -\frac{1}{\kappa} P_{(1)j}, \tag{4.168}$$

$$\left\{ P_{(1)j}, x^k_{(2)} \right\} = \delta^k_j \left(1 - e^{-P_{(1)0}/\kappa} \right), \quad \left\{ P_{(2)\mu}, x^\nu_{(1)} \right\} = 0, \tag{4.169}$$

$$\left\{ x^0_{(1)}, x^j_{(1)} \right\} = \left\{ x^0_{(1)}, x^j_{(2)} \right\} = \left\{ x^0_{(2)}, x^j_{(1)} \right\} = -\frac{1}{\kappa} x^j_{(1)}, \tag{4.170}$$

$$\left\{ x^0_{(2)}, x^j_{(2)} \right\} = -\frac{1}{\kappa} x^j_{(2)}, \quad \left\{ x^j_{(a)}, x^k_{(a)} \right\} = 0. \tag{4.171}$$

The Hamiltonian of the system is

$$\mathscr{H} = \mathscr{L} - \sum_i \tilde{x}^\mu \dot{p}_\mu = \sum_i \lambda_{(a)} \left(\mathscr{C}_{(a)} - m^2_{(a)} \right), \tag{4.172}$$

and it generates the evolution in time τ. Indeed, using the Poisson brackets above, one can verify that Eqs. (4.164) and (4.165) can be rewritten as

$$\dot{p}_{(a)\mu} = \left\{ \mathscr{H}, p_\mu \right\} = 0, \qquad \dot{x}^\mu_{(a)} = \left\{ \mathscr{H}, x^\mu \right\}. \tag{4.173}$$

This completes our discussion of the kinematical properties of the deformed two-particle system and lets us now turn to its symmetries.

4.7.1 Translations

Let us now turn to the symmetries of the system of two deformed particles described by the Lagrangian (4.163). We start with rigid translations and the associated notion of total momentum. This will provide us with the intuition concerning the physical meaning of the Hopf algebra structure called the coproduct (see next chapter).

Clearly, the total momentum of the system of particles is a conserved quantity. But as we have seen in (4.164), the momenta of individual particles composing the two-particle system are independently conserved. In the standard, undeformed theory, there is a little alternative to defining the total momentum as a sum of the individual momenta. The reason is, as always, that in order to go beyond linearity, one needs a momentum scale which is not available in the undeformed theory. When the deformation is present, the situation changes dramatically and there is no obvious candidate for the total momentum of a multi-particle system. The conservation does not help here, because, since the individual particles' momenta are conserved, any (even non-linear) combination of them is conserved as well. It follows that the requirement for the total momentum to be conserved does not single out a unique generator for translations. In principle, the conserved charges, and hence the generators of symmetries, associated with translations, can be chosen arbitrarily. There must be therefore some other property, besides conservation, that can be used to define what the translation generators are. A natural candidate for the defining condition is the requirement that the total momentum generate 'rigid translations' of the two-particle system, i.e., to be such that the relative position of the particles does not change, in other words, the coordinates of both particles translate by the same amount.

This rigidity requirement is based on the following physical intuition. Put the particles into an imaginary box. The total momentum generates a translation of the box with particles inside by ξ^μ. If we look inside the box now, we could demand that each individual particle is being translated, by its own momentum, in exactly the same way.

It is not guaranteed by any means that such total momentum exists in general and it is a remarkable property of the model we are considering here, directly related to the non-symmetric form of the Lagrangian (4.159), that the total momentum generating rigid transformations does exist and moreover is exactly the total momentum defined by the $AN(3)$ group product (see Eq. (4.184) below).

To see how this argument works, let us revisit the Lagrangian (4.161) and express it in terms of the relative position of the particles $x^\mu_{(-)} = 1/2 \left(x^\mu_{(2)} - x^\mu_{(1)} \right)$ and the average position $x^\mu_{(+)} = 1/2 \left(x^\mu_{(2)} + x^\mu_{(1)} \right)$. It follows from our discussion above that under rigid translations, with translation parameter $\xi^\mu = \left(\xi^0, \boldsymbol{\xi} \right)$, such that $x'^\mu_{(a)} = x^\mu_{(a)} + \delta_\xi x^\mu_{(a)}$, the relative position does not change: $\delta_\xi x^\mu_{(-)} = 0$. Therefore, the variation of the Lagrangian under infinitesimal translations will be proportional to $\delta_\xi x^\mu_{(+)}$. Notice first that, after rearranging the coordinates, the variation of the kinetic term is

$$\delta_\xi \mathcal{L}_{1\oplus2}^{kin} = \delta_\xi x_{(+)}^0 \left(\dot{p}_{(1)0} + \dot{p}_{(2)0} \right)$$
$$+ \delta_\xi x_{(+)}^j \left[\dot{p}_{(1)j} + e^{-P_{(1)0}/\kappa} \dot{p}_{(2)j} - \frac{1}{\kappa} e^{-P_{(1)0}/\kappa} p_{(2)j} \dot{p}_{(1)0} \right.$$
$$\left. + \frac{1}{\kappa} \left(p_{(1)j} + e^{-P_{(1)0}/\kappa} p_{(2)j} \right) \left(\dot{p}_{(1)0} + \dot{p}_{(2)0} \right) \right] . \tag{4.174}$$

We want the coordinate variation to be $\delta_\xi x_{(+)}^\mu = -\xi^\mu + O\left(1/\kappa\right)$, with constant parameters ξ^μ, in order to recover the standard translation in the limit $1/\kappa \to 0$. Let us first consider terms in the variation proportional to ξ^0. It is easy to see from (4.174) that imposing

$$\delta_{\xi^0} x_{(+)}^0 = -\xi^0, \qquad \delta_{\xi^0} x_{(+)}^j = 0, \tag{4.175}$$

we get

$$\delta_{\xi^0} \mathcal{L}_{1\oplus2}^{kin} = -\xi^0 \frac{d}{d\tau} \left(p_{(1)0} + p_{(2)0} \right), \tag{4.176}$$

so that the zeroth component of the conserved total momentum is

$$p_0^{tot} = p_{(1)0} + p_{(2)0}. \tag{4.177}$$

It is a bit less trivial to find the spacial component of the conserved total momentum. From (4.174), one can notice that the terms proportional to $\delta_\xi x_{(+)}^j$ do not add up to a total derivative. However, one can verify that by setting the variation parametrized by $\boldsymbol{\xi}$ to be

$$\delta_\xi x_{(+)}^0 = \frac{1}{\kappa} \xi^j \left(p_{(1)j} + e^{-P_{(1)0}/\kappa} p_{(2)j} \right), \qquad \delta_\xi x_{(+)}^j = -\xi^j, \tag{4.178}$$

we get

$$\delta_\xi \mathcal{L}_{1\oplus2}^{kin} = -\xi^j \frac{d}{d\tau} \left(p_{(1)j} + e^{-P_{(1)0}/\kappa} p_{(2)j} \right), \tag{4.179}$$

so that the spacial component of the total momentum is

$$p_j^{tot} = p_{(1)j} + e^{-P_{(1)0}/\kappa} p_{(2)j}. \tag{4.180}$$

One can check using (4.167)–(4.169) that the total momentum p_μ^{tot} (4.177) (4.180) generates the translations (4.175) (4.178) by Poisson brackets, its action on the single-particle coordinates being

$$\delta_\xi x_{(a)}^\mu = -\left\{ \xi^\nu p_\nu^{tot}, x_{(a)}^\mu \right\},$$
$$\delta_\xi x_{(1)}^0 = \delta_\xi x_{(2)}^0 = -\xi^0 + \frac{1}{\kappa} \boldsymbol{\xi} \cdot \mathbf{p}^{tot}, \qquad \delta_\xi x_{(1)}^j = \delta_\xi x_{(2)}^j = -\xi^j. \tag{4.181}$$

Both particles are translated by the same amount in agreement with what we assumed about rigid translations. Notice also that the total momentum generating rigid translations can be re-expressed as a deformed summation law for the single-particle momenta

$$p_\mu^{tot} = \left(p_{(1)} \oplus p_{(2)} \right)_\mu , \tag{4.182}$$

with

$$\left(p_{(1)} \oplus p_{(2)} \right)_0 = p_{(1)0} + p_{(2)0}, \quad \left(p_{(1)} \oplus p_{(2)} \right)_j = p_{(1)j} + e^{-p_{(1)0}/\kappa} p_{(2)j} . \tag{4.183}$$

As we will see in the next section, this momentum composition law is directly related to the κ-Poincaré Hopf algebra structure called the coproduct.

This deformed summation law, as it turns out, can be associated with the product of the two-particle group-valued momenta as

$$\Pi^{tot} \left(p_{(1)} \oplus p_{(2)} \right) = \Pi_{(1)} \left(p_{(1)} \right) \Pi_{(2)} \left(p_{(2)} \right) . \tag{4.184}$$

4.7.2   Lorentz Symmetry

As we saw in the previous subsection, our action for a κ-deformed two-particle system is characterized by its total momentum, which in turn implies that the translations, generated by the total momentum of the system, are rigid, so that the locality of a single process is preserved under translation. We want now to study if a similar property is fulfilled by the remaining spacetime transformations, boosts, and rotations.

The mass-shell relation for each particle of the two-particle system is the κ-Poincaré mass Casimir (4.162). The single-particle boost and rotation charges (generators) compatible with this choice of Casimir satisfy the κ-Poincaré algebra, first derived in this form in [16]

$$\left\{ \mathcal{N}_{(a)j}, P_{(a)0} \right\} = P_{(a)j} ,$$

$$\left\{ \mathcal{N}_{(a)j}, P_{(a)k} \right\} = \delta_{jk} \left(\frac{\kappa}{2} \left(1 - e^{-2P_{(a)0}/\kappa} \right) + \frac{1}{2\kappa} \mathbf{p}_{(a)}^2 \right) - \frac{1}{\kappa} P_{(a)j} P_{(a)k} ,$$

$$\left\{ R_{(a)j}, P_{(a)0} \right\} = 0 ,$$

$$\left\{ R_{(a)j}, P_{(a)k} \right\} = \varepsilon_{jkl} P_{(a)l} ,$$

$$\left\{ R_{(a)j}, R_{(a)k} \right\} = \varepsilon_{jkl} R_{(a)l} ,$$

$$\left\{ R_{(a)j}, \mathcal{N}_{(a)k} \right\} = \varepsilon_{jkl} \mathcal{N}_{(a)l} ,$$

$$\left\{ \mathcal{N}_{(a)j}, \mathcal{N}_{(a)k} \right\} = -\varepsilon_{jkl} R_{(a)l} , \tag{4.185}$$

and they are conserved charges since from (4.172) it follows

$$\dot{\mathcal{N}}_{(a)j} = \left\{ \mathcal{H}, \mathcal{N}_{(a)j} \right\} = 0, \qquad \dot{R}_{(a)j} = \left\{ \mathcal{H}, R_{(a)j} \right\} = 0. \tag{4.186}$$

Using relations (4.167), (4.169), and (4.171), one finds the representation of the single-particle boost and rotation generators in terms of phase space variables to be

$$\mathcal{N}_{(1)j} = -p_{(1)j}x_{(1)}^0 - \left(\frac{\kappa}{2}\left(1 - e^{-2p_{(1)0}/\kappa}\right) + \frac{1}{2\kappa}\mathbf{p}_{(1)}^2\right)x_{(1)}^j, \tag{4.187}$$

$$\mathcal{N}_{(2)j} = -p_{(2)j}x_{(2)}^0 - \left(\frac{\kappa}{2}\left(1 - e^{-2p_{(2)0}/\kappa}\right) + \frac{1}{2\kappa}\mathbf{p}_{(2)}^2\right)\left(x_{(2)}^j - \left(1 - e^{-p_{(1)0}/\kappa}\right)x_{(1)}^j\right)$$

$$- \frac{1}{\kappa}p_{(2)j}\mathbf{p}_{(1)} \cdot \mathbf{x}_{(1)}, \tag{4.188}$$

$$R_{(1)j} = \varepsilon_{jkl}p_{(1)k}x_{(1)}^l,$$

$$R_{(2)j} = \varepsilon_{jkl}p_{(2)k}\left(x_{(2)}^l - \left(1 - e^{-p_{(1)0}/\kappa}\right)x_{(1)}^l\right). \tag{4.189}$$

To derive the expression for the total boost and rotation generators, we notice that the two-particle system is seen by an observer with not sufficient 'resolution power' as a single system carrying the momentum $p_\mu^{(tot)}$. It follows that the total momentum should transform, with respect to the symmetries generated by total boost $\mathcal{N}_j^{(tot)}$ and rotation $R_j^{(tot)}$, in exactly the same way as the momenta $p_{(a)\mu}$ transform under the symmetries generated by the single-particle boost $\mathcal{N}_{(a)j}$ and rotation $R_{(a)j}$. Thus, we want the total boost and total rotation generators to satisfy the property

$$\left\{\mathcal{N}_j^{tot}, p_\mu^{tot}\right\} = \left\{\mathcal{N}_{(a)j}, p_{(a)\mu}\right\}\Big|_{\substack{p_{(a)}\to p^{tot}, \\ \mathcal{N}_{(a)}\to\mathcal{N}^{tot}}}$$

$$\left\{R_j^{tot}, p_\mu^{tot}\right\} = \left\{R_{(a)j}, p_{(a)\mu}\right\}\Big|_{\substack{p_{(a)}\to p^{tot}. \\ R_{(a)}\to R^{tot}}} \tag{4.190}$$

One finds that the total boost and rotation generators satisfying this relation have the expression

$$\mathcal{N}_j^{tot} = \mathcal{N}_{(1)j} + e^{-p_{(1)0}/\kappa}\mathcal{N}_{(2)j} + \frac{1}{\kappa}\varepsilon_{jkl}p_{(1)k}R_{(2)l}, \qquad R_j^{tot} = R_{(1)j} + R_{(2)j}, \tag{4.191}$$

which is exactly the expression that follows from the κ-Poincaré coproducts [16]. We see again the remarkable property that the expressions for the generators of total momentum, boost, and rotation deduced in our classical model on the basic physical premises reproduce the coproduct structure of the κ-Poincaré Hopf algebra. One can verify that the property for the total generators to transform in the same way as the single-particle generators, i.e., for the total generators to satisfy the same algebra of the single-particle ones, is satisfied for all the set of spacetime symmetries: calling $\mathcal{G}_{(a)\mu}$ the generic element of the set of single-particle charge/generators (momenta, boosts, and rotations), and $\mathcal{G}_\mu^{tot}$ their two-particle composite version, one can verify with the help of relations (4.185), that the following property is satisfied:

$$\left\{\mathcal{G}_\mu^{tot}, \mathcal{G}_\nu^{tot}\right\} = \left\{\mathcal{G}_{(a)\mu}, \mathcal{G}_{(a)\nu}\right\}\Big|_{\mathcal{G}_{(a)}\to\mathcal{G}^{tot}}. \tag{4.192}$$

We now turn to examine the behavior of coordinate changes under the action of the total boost and rotation generators defined in the previous subsection. Under the action of a composite boost and rotation, the spacetime coordinates change, respectively, as

$$
x'^{\mu}_{(a)} = x^{\mu}_{(a)} + \delta_{\lambda} x^{\mu}_{(a)} = x^{\mu}_{(a)} - \lambda \cdot \left\{ \mathcal{N}^{tot}, x^{\mu}_{(a)} \right\} ,
$$
$$
x'^{\mu}_{(a)} = x^{\mu}_{(a)} + \delta_{\theta} x^{\mu}_{(a)} = x^{\mu}_{(a)} - \boldsymbol{\theta} \cdot \left\{ \mathbf{R}^{tot}, x^{\mu}_{(a)} \right\} ,
\tag{4.193}
$$

where λ^j and θ^j are the boost and rotation parameters. The system of particle transforms under the action of the composite boost and rotation (4.191) as

$$
\left\{ \mathcal{N}^{tot}_j, x^0_{(1)} \right\} = -x^j_{(1)} - \frac{1}{\kappa} \mathcal{N}^{tot}_j , \qquad \left\{ \mathcal{N}^{tot}_j, x^k_{(1)} \right\} = -\delta_{jk} x^0_{(1)} + \frac{1}{\kappa} \varepsilon_{jkl} R^{tot}_l ,
\tag{4.194}
$$

$$
\left\{ \mathcal{N}^{tot}_j, x^0_{(2)} \right\} = -e^{-P_{(1)0}/\kappa} x^j_{(2)} - \left(1 - e^{-P_{(1)0}/\kappa} \right) x^j_{(1)} - \frac{1}{\kappa} \mathcal{N}^{tot}_j ,
\tag{4.195}
$$

$$
\left\{ \mathcal{N}^{tot}_j, x^k_{(2)} \right\} = -\delta_{jk} \left(e^{-P_{(1)0}/\kappa} x^0_{(2)} + \left(1 - e^{-P_{(1)0}/\kappa} \right) x^0_{(1)} - \frac{1}{\kappa} \mathbf{P}_{(1)} \cdot \left(\mathbf{x}_{(2)} - \mathbf{x}_{(1)} \right) \right)
$$
$$
+ \frac{1}{\kappa} \varepsilon_{jkl} R^{tot}_l - \frac{1}{\kappa} P_{(1)j} \cdot \left(x^k_{(2)} - x^k_{(1)} \right) ,
\tag{4.196}
$$

$$
\left\{ R^{tot}_j, x^0_{(1)} \right\} = \left\{ R^{tot}_j, x^0_{(2)} \right\} = 0 ,
$$

$$
\left\{ R^{tot}_j, x^k_{(1)} \right\} = \varepsilon_{jkl} x^l_{(1)} , \qquad \left\{ R^{tot}_j, x^k_{(2)} \right\} = \varepsilon_{jkl} x^l_{(2)} .
\tag{4.197}
$$

One can verify that the action of boosts and rotations on the composite system satisfies the property

$$
\left(\left\{ \mathcal{N}^{tot}, x^{\mu}_{(1)} \right\} - \left\{ \mathcal{N}^{tot}, x^{\mu}_{(2)} \right\} \right) \Big|_{x_{(1)} = x_{(2)}} = 0 ,
$$
$$
\left(\left\{ R^{tot}, x^{\mu}_{(1)} \right\} - \left\{ R^{tot}, x^{\mu}_{(2)} \right\} \right) \Big|_{x_{(1)} = x_{(2)}} = 0 .
\tag{4.198}
$$

Notice that the interaction point, for a first observer that describes the interaction as local, is identified by the condition ($x_{(1)} = x_{(2)}$), i.e., the coinciding endpoint of the two particles' worldlines. Then, considering from Eq. (4.193) that the change on the coordinates is proportional to their Poisson brackets with the total generators, Eq. (4.198) implies that if the interaction point is local for a first observer, it remains local also for a boosted or rotated observer:

$$
\text{if } x_{(1)} = x_{(2)} \text{ then } x'_{(1)} = x'_{(2)} .
\tag{4.199}
$$

Thus, as it was in the case of translations generated by the total momentum, the locality of a distant process is preserved also by the total boosts and rotations defined in this section.

References

1. Arzano, M., Nettel, F.: Deformed phase spaces with group valued momenta. Phys. Rev. D **94**(8), 085004 (2016). arXiv:1602.05788 [hep-th]
2. Nakahara, M.: Geometry, Topology and Physics. IOP Publishing (2003)
3. Abraham, R., Marsden, J.E.: Foundation of Mechanics. Benjamin/Cummings Publishing Company (1978)
4. Chari, V., Pressley, A.: A Guide to Quantum Groups. Cambridge University Press (1998)
5. Alekseev, A.Yu., Malkin, A.Z.: Commun. Math. Phys. **162**, 147 (1994)
6. Kosmann-Schwarzbach, Y.: Lecture Notes in Physics **495**, 104 (1997). In Proceedings of the CIMPA School Pondicherry University, India, 8–26 January 1996. Springer, Berlin, Heidelberg (1997)
7. Arzano, M., Trzesniewski, T.: Diffusion on κ-Minkowski space. Phys. Rev. D **89**(12), 124024 (2014). arXiv:1404.4762 [hep-th]
8. Fuchs, J., Schweigert, C.: Symmetries, Lie Algebras and Representations. Cambridge University Press (1997)
9. Bonzom, V., Dupuis, M., Girelli, F., Livine, E.R.: arXiv:1402.2323 [gr-qc]
10. Arzano, M., Latini, D., Lotito, M.: SIGMA **10**, 079 (2014)
11. Amelino-Camelia, G., Freidel, L., Kowalski-Glikman, J., Smolin, L.: The principle of relative locality. Phys. Rev. D **84**, 084010 (2011). arXiv:1101.0931 [hep-th]
12. Cianfrani, F., Kowalski-Glikman, J., Rosati, G.: Phys. Rev. D **89**(4), 044039 (2014). https://doi.org/10.1103/PhysRevD.89.044039, arXiv:1401.2057 [gr-qc]
13. Lukierski, J., Ruegg, H., Nowicki, A., Tolstoi, V.N.: Q deformation of Poincare algebra. Phys. Lett. B **264**, 331 (1991)
14. Lukierski, J., Nowicki, A., Ruegg, H.: New quantum Poincare algebra and k deformed field theory. Phys. Lett. B **293**, 344 (1992)
15. Lukierski, J., Ruegg, H., Zakrzewski, W.J.: Classical quantum mechanics of free kappa relativistic systems. Ann. Phys. **243**, 90 (1995). arXiv:hep-th/9312153
16. Majid, S., Ruegg, H.: Bicrossproduct structure of kappa Poincare group and noncommutative geometry. Phys. Lett. B **334**, 348 (1994). arXiv:hep-th/9405107
17. Freidel, L., Kowalski-Glikman, J., Nowak, S.: Field theory on kappa-Minkowski space revisited: Noether charges and breaking of Lorentz symmetry. Int. J. Mod. Phys. A **23**, 2687 (2008). arXiv:0706.3658 [hep-th]
18. Gubitosi, G., Mercati, F.: Relative Locality in κ-Poincaré. arXiv:1106.5710 [gr-qc]
19. Borowiec, A., Pachol, A.: κ-Minkowski spacetimes and DSR algebras: fresh look and old problems. SIGMA **6**, 086 (2010). arXiv:1005.4429 [math-ph]
20. Amelino-Camelia, G., Majid, S.: Waves on noncommutative space-time and gamma-ray bursts. Int. J. Mod. Phys. A **15**, 4301 (2000). arXiv:hep-th/9907110
21. Kirillov, A.A.: Elements of the Theory of Representations. Springer, Berlin (1976)
22. Amelino-Camelia, G., Arzano, M., Kowalski-Glikman, J., Rosati, G., Trevisan, G.: Relative-locality distant observers and the phenomenology of momentum-space geometry. Class. Quant. Grav. **29**, 075007 (2012). arXiv:1107.1724 [hep-th]
23. Daszkiewicz, M., Imilkowska, K., Kowalski-Glikman, J.: Velocity of particles in doubly special relativity. Phys. Lett. A **323**, 345 (2004). arXiv:hep-th/0304027
24. Kowalski-Glikman, J., Rosati, G.: Multi-particle systems in κ-Poincaré inspired by 2+1D gravity. Phys. Rev. D **91**(8), 084061 (2015). https://doi.org/10.1103/PhysRevD.91.084061, arXiv:1412.0493 [hep-th]

Hopf Algebra Relativistic Symmetries: The κ-Poincaré Algebra

5

In the first part of this book, we argued that if one takes into account self-gravity effects, a particle becomes dressed with topological degrees of freedom of its own gravitational field so that its kinematics becomes deformed. This results in the energy and momentum of the particle being described by elements of a non-abelian Lie group. In this chapter, we turn to the mathematics describing this deformation.

The first step will be the introduction of some key notions in the theory of Hopf algebras and their relevance in physical applications. We then specialize to the four-dimensional example of Hopf algebra deformations of relativistic symmetries given by the κ-Poincaré algebra which will be the subject of the second part of these notes in which we develop some aspects of the associated non-commutative quantum field theory.

5.1   Hopf Algebra Structures in Quantum Theory

Perhaps the deepest difference between the mathematical structure of classical mechanics and quantum mechanics is the way composite systems are described.

In the classical theory, a system comprised of, say, two particles is described by the cartesian product of their phase spaces $\Gamma = \Gamma_1 \times \Gamma_2$, i.e., states of this composite system are described by ordered pairs of the states of the components of the system. In quantum theory, the states of an elementary system are vectors in a complex Hilbert space $\mathscr{H}$. The correct description of a composite system is now given by the tensor product of its components. This can be easily understood in terms of the wave-function description of the individual subsystems. Indeed, if $\psi_1(\mathbf{x}) \in \mathscr{H}_1$ and $\psi_2(\mathbf{y}) \in \mathscr{H}_2$ are the wave-functions describing the states of two-particles, the probabilistic interpretation of these states requires that $\int |\psi_1(\mathbf{x})|^2 d\mathbf{x} = \int |\psi_2(\mathbf{y})|^2 d\mathbf{y} = 1$ and thus $\mathscr{H}_1 = \mathscr{H}_2 = \mathscr{L}^2(\mathbb{R}^3)$. The state of the composite system will be described

© Springer-Verlag GmbH Germany, part of Springer Nature 2021
M. Arzano and J. Kowalski-Glikman, *Deformations of Spacetime Symmetries*,
Lecture Notes in Physics 986,
https://doi.org/10.1007/978-3-662-63097-6_5

by a function $\psi(\mathbf{x}, \mathbf{y})$ such that $\int |\psi(\mathbf{x}, \mathbf{y})|^2 d\mathbf{x}\, d\mathbf{y} = 1$. This is a square integrable function defined on two copies of $\mathbb{R}^3$, i.e., $\psi(\mathbf{x}, \mathbf{y}) \in \mathscr{L}^2(\mathbb{R}^3 \times \mathbb{R}^3)$. Elements of such space are functions of six variables and it is obvious that not all such functions can be written as products of two functions of three variables in $\mathscr{L}^2(\mathbb{R})^3 \times \mathscr{L}^2(\mathbb{R}^3)$. In fact, it can be proven that $\mathscr{L}^2(\mathbb{R}^3 \times \mathbb{R}^3)$ is isomorphic to the tensor product of the single particle Hilbert spaces $\mathscr{L}^2(\mathbb{R}^3 \times \mathbb{R}^3) \equiv \mathscr{L}^2(\mathbb{R}^3) \otimes \mathscr{L}^2(\mathbb{R}^3)$ [1]. Thus, states of composite systems in quantum mechanics are described by elements of *tensor products of the Hilbert spaces of their components*. This simple but fundamental fact is at the basis of typical quantum phenomena, for example, *quantum entanglement*, which make quantum systems so radically different from their classical counterparts [2].

In relativistic quantum theory, invariance under translations and Lorentz transformations requires that the states describing elementary particles should carry a unitary irreducible representation of the Poincaré group. For the simplest example of a real scalar field, we have a 'one-particle' Hilbert space $\mathscr{H}$ whose elements can be thought of positive energy solutions of the Klein–Gordon equation or, equivalently, as complex functions on the mass-shell in four-momentum space and whose elements we denote as kets labelled by the linear momentum carried by the particle $|\mathbf{k}\rangle \in \mathscr{H}$.

As seen above, the states of systems consisting of many particles are described by elements of tensor products of one-particle Hilbert spaces. The quanta of the scalar field are bosons and when we deal with systems of identical particles, their indistinguishability requires that we represent their states with combinations of tensor products of their Hilbert spaces which are symmetric under exchange of their labels. In other words, such states must carry a representation of the symmetric group. For our scalar field 'n-particle states' will be represented by symmetrized n-tensor products of the one-particle space $\mathscr{H}$, i.e., their Hilbert space can be written as

$$S_n \mathscr{H}^n = \frac{1}{n!} \sum_{\sigma \in P_n} \sigma(\mathscr{H}^{\otimes n}) . \tag{5.1}$$

where σ is a permutation in the permutation group of n-elements P_n and $\mathscr{H}^{\otimes n}$ is the n-fold tensor product of n-copies of $\mathscr{H}$. The full Hilbert space of the theory is given by a direct sum of these n-particle Hilbert spaces and is called the *Fock space*

$$\mathscr{F}(\mathscr{H}) = \bigoplus_{n=0}^{\infty} S_n \mathscr{H}^n , \tag{5.2}$$

where $\mathscr{H}^0 = \mathbb{C}$.

Now let us switch our attention from states to *observables*. For a single particle system, these are self-adjoint operators on the Hilbert space $\mathscr{H}$. The simplest example of observables are the generators of transformations of the Poincaré group and in particular of spacetime translations P_μ which we use to characterize the energy and momentum of the particles. In general, given a 'one-particle' observable $\mathscr{O}$ its action on multi-particle states, i.e., on a generic element of the Fock space, is given by the *second quantized* operator [3]

$$d\Gamma(\mathcal{O}) \equiv 1 + \mathcal{O} + (\mathcal{O} \otimes 1 + 1 \otimes \mathcal{O}) + (\mathcal{O} \otimes 1 \otimes 1 + 1 \otimes \mathcal{O} \otimes 1 + 1 \otimes 1 \otimes \mathcal{O}) + \cdots \tag{5.3}$$

where 1 is the identity operator. The expression (5.3) is simply telling us that the operator $d\Gamma(\mathcal{O})$ acts on multi-particle states as a derivative, i.e., following the Leibnitz rule. However, this quite obvious expression has an important mathematical significance. The expression above can, in fact, be rewritten in terms of an operation on $\mathcal{O}$ which we call the *coproduct*

$$\Delta\mathcal{O} = \mathcal{O} \otimes 1 + 1 \otimes \mathcal{O} \tag{5.4}$$

as

$$d\Gamma(\mathcal{O}) \equiv 1 + \mathcal{O} + \Delta\mathcal{O} + \Delta_2\mathcal{O} + \cdots + \Delta_n\mathcal{O} + \cdots \tag{5.5}$$

where

$$\Delta_n\mathcal{O} = (\Delta \otimes 1) \circ \Delta_{n-1} , \qquad n \geq 2 \tag{5.6}$$

with $\Delta_1 \equiv \Delta$. The coproduct (5.4) contains information on how the operator $\mathcal{O}$ is 'shared-out' when acting on a tensor product of Hilbert spaces $\mathcal{H} \otimes \mathcal{H}$. This operation is somewhat complementary to that of multiplication of two observables which takes input from two copies of the algebra and produces a result in one copy of it and this justifies the term 'coproduct'. Let us focus on the specific example of the observables P_i associated to the linear momentum of the particle. On one-particle states, we have

$$P_i|\mathbf{k}\rangle = k_i|\mathbf{k}\rangle \tag{5.7}$$

where k_i is the i-th component of the vector $\mathbf{k}$. The coproduct

$$\Delta P_i = P_i \otimes 1 + 1 \otimes P_i \tag{5.8}$$

tells us how the observable linear momentum is acting on a two-particle state

$$|\mathbf{k}\,\mathbf{l}\rangle \equiv \frac{1}{\sqrt{2}}\left(|\mathbf{k}\rangle \otimes |\mathbf{l}\rangle + |\mathbf{k}\rangle \otimes |\mathbf{l}\rangle\right) \tag{5.9}$$

in particular

$$\Delta P_i|\mathbf{k}\,\mathbf{l}\rangle = (k_i + l_i)|\mathbf{k}\,\mathbf{l}\rangle , \tag{5.10}$$

giving us the total linear momentum of the composite system. In other words, the coproduct encodes the crucial property of *additivity of quantum numbers*. Thus, by tacitly assuming this property, we are introducing an additional structure on the set of observables which goes beyond that of a unital associative algebra.

Before proceeding with the formalization of such new structure, we need another ingredient coming this time from the action of observables on the bras $\langle k|$, i.e., on elements of the dual to the one-particle Hilbert space $\mathscr{H}^*$. We first discuss this feature at a more formal level and then give a physical interpretation. By definition, elements of $\mathscr{H}^*$ are continuous linear maps from $\mathscr{H}$ to $\mathbb{C}$ and indeed writing the inner product on $\mathscr{H}$ as $\langle \mathbf{k}'|\mathbf{k}\rangle$, we see that, indeed, the bra $\langle \mathbf{k}|$ can be seen as an element of $\mathscr{H}^*$. Now if $\mathscr{H}$ carries a representation of a Lie algebra, like the case of a relativistic one-particle Hilbert space which carries a representation of the Poincaré algebra, there exists a corresponding *dual* representation of the same Lie algebra on $\mathscr{H}^*$ (see, e.g., [4]). Let us take, for example, the action of spatial translation generators on one-particle states (5.7)

$$P_i|\mathbf{k}\rangle = k_i|\mathbf{k}\rangle \,, \tag{5.11}$$

then the action of P_i on a vector $\langle \mathbf{k}'| \in \mathscr{H}^*$ is such that

$$(P_i\langle \mathbf{k}'|)|\mathbf{k}\rangle = -\langle \mathbf{k}'|(P_i|\mathbf{k}\rangle) \,. \tag{5.12}$$

In other words, the dual representations defines an action *from the right* of the translations generators on bras given by

$$P_i\langle \mathbf{k}| = -k_i\langle \mathbf{k}| \,. \tag{5.13}$$

This is to be contrasted with the action from the left which is simply given by taking the hermitian adjoint of (5.7)

$$\langle \mathbf{k}|P_i \equiv (P_i|\mathbf{k}\rangle)^\dagger = \langle \mathbf{k}|k_i \,. \tag{5.14}$$

In light of this observation, we can rewrite (5.13) in terms of a map connecting the left and right action on dual states

$$P_i\langle \mathbf{k}| = -k_i\langle \mathbf{k}| = \langle \mathbf{k}|(-k_i) \equiv \langle \mathbf{k}|S(P_i) \tag{5.15}$$

where the *antipode map* $S(P_i)$ is given by

$$S(P_i) = -P_i \,. \tag{5.16}$$

In order to get some physical intuition on the meaning of this operation, let us recall that starting with the one-particle Hilbert space $\mathscr{H}$, the space describing antiparticles is given by the complex conjugate space $\bar{\mathscr{H}}$. The latter is comprised of the same vector space on which $\mathscr{H}$ is built but with the operation of multiplication by a complex number replaced by multiplication by the *complex conjugate* number. For a scalar field in Minkowski spacetime, this translates into functions on the positive mass-shell (one-particle states) getting multiplied by a complex number, while functions on the negative mass-shell (one-antiparticle states) getting multiplied by its complex conjugate and this ensures that, for a complex field, antiparticle states have positive energy. It turns out [5] that the complex conjugate Hilbert space $\bar{\mathscr{H}}$ is isomorphic to the dual Hilbert space $\mathscr{H}^*$, and thus for a complex scalar field the bras $\langle k|$ can be seen as representatives of antiparticle states. For a real scalar field, one can identify

$\langle -k| \equiv |k\rangle$ which is nothing but a reflection of the familiar reality condition for the field $\bar{\phi}(-k) = \phi(k)$ which provides a natural isomorphism between $\mathcal{H}$ and $\bar{\mathcal{H}}$ [6]. We thus see that the antipode map introduced above is associated with the way observables act on antiparticle states. We will return on this point when we will discuss in more detail κ-deformed quantum field in Part III of this book.

What we would like to point out here is that the mere existence of multi-particle and antiparticle states in relativistic quantum theory naturally leads to the introduction of additional structures on the algebra of observables. To fix our ideas, let us focus again on observables associated with the isometries of Minkowski spacetime, i.e., to the generators of the Poincaré algebra. The algebra of observables associated with such generators is an *associative algebra* [7] generated from the Poincaré Lie algebra known as the *universal enveloping algebra*. Let us recall how such algebra is constructed (for further mathematical details see [8]). Let $V_{\mathfrak{iso}(3,1)}$ be the vector space spanned by the generators of the Poincaré Lie algebra $\mathfrak{iso}(3,1)$. The tensor algebra of $V_{\mathfrak{iso}(3,1)}$ is the algebra obtained by considering the direct sum of all possible tensor products $V^{(n)}_{\mathfrak{iso}(3,1)} = \underbrace{V_{\mathfrak{iso}(3,1)} \otimes V_{\mathfrak{iso}(3,1)} \otimes ... \otimes V_{\mathfrak{iso}(3,1)}}_{n-\text{times}}$ of such vector space

$$\mathcal{T}(V_{\mathfrak{iso}(3,1)}) \equiv \bigoplus_{n=0}^{\infty} (V^{(n)}_{\mathfrak{iso}(3,1)}), \tag{5.17}$$

where $V^{(0)} = \mathbb{C}$, equipped with a natural product given by the tensor product $\otimes$ between two of its elements. Loosely speaking, we can think of elements of this space as tensor polynomials in the elements of $V_{\mathfrak{iso}(3,1)}$ representing possible combinations of measurements of the observables associated with Minkowski spacetime isometries (like energy, angular momentum, etc.). This algebra, however, knows nothing about the commutators of the original algebra $\mathfrak{iso}(3,1)$ since for its construction we have just used the vector space $V_{\mathfrak{iso}(3,1)}$. Indeed there is a redundancy in the way $\mathcal{T}(V_{\mathfrak{iso}(3,1)})$ describes possible measurements on the system under consideration (e.g., an elementary relativistic particle) since, for example, given two distinct elements $X_1, X_2 \in V_{\mathfrak{iso}(3,1)}$ the elements

$$X_1 \otimes X_2 - X_2 \otimes X_1 \in V^{(2)}_{\mathfrak{iso}(3,1)}, \tag{5.18}$$

and

$$[X_1, X_2] \in V^{(1)}_{\mathfrak{iso}(3,1)}, \tag{5.19}$$

should represent the same measurement. Thus, any two generic elements of $\mathcal{T}(V_{\mathfrak{iso}(3,1)})$ like

$$X_1 \otimes ... \otimes X_i \otimes X_{i+1} \otimes ... \otimes X_n - X_1 \otimes ... \otimes X_{i+1} \otimes X_i \otimes ... \otimes X_n \tag{5.20}$$

and

$$X_1 \otimes ... \otimes [X_i, X_{i+1}] \otimes ... \otimes X_n \tag{5.21}$$

should be identified or, in other words, elements like $X_1 \otimes \ldots \otimes X_i \otimes X_{i+1} \otimes \ldots \otimes X_n - X_1 \otimes \ldots \otimes X_{i+1} \otimes X_i \otimes \ldots \otimes X_n - X_1 \otimes \ldots \otimes [X_i, X_{i+1}] \otimes \ldots \otimes X_n$ should be set to zero. At a formal level, this can be accomplished by considering the quotient space of the algebra $\mathcal{T}(V_{\mathrm{iso}(3,1)})$ by the set of all these elements. The latter is defined as *the smallest two-sided ideal* $\mathcal{I}$, i.e., the smallest set such that for any $X \in \mathcal{T}(V_{\mathrm{iso}(3,1)})$ one has $X \otimes \mathcal{I} \subseteq \mathcal{I}$ and $\mathcal{I} \otimes X \subseteq \mathcal{I}$, containing all elements of the type $X_1 \otimes X_2 - X_2 \otimes X_1 - [X_1, X_2]$. The quotient

$$U(V_{\mathrm{iso}(3,1)}) = \mathcal{T}(V_{\mathrm{iso}(3,1)})/\mathcal{I} , \tag{5.22}$$

namely the set of equivalence classes $[X]$ of elements of $\mathcal{T}(V_{\mathrm{iso}(3,1)})$ with respect to $\mathcal{I}$, the classes of elements of $\mathcal{T}(V_{\mathrm{iso}(3,1)})$ which differ only by an element of $\mathcal{I}$ (which we want to set to zero), is known as the *universal enveloping algebra* of the Lie algebra $\mathrm{iso}(3,1)$ (a construction obviously valid for any Lie algebra). The (associative) product in $V_{\mathrm{iso}(3,1)}$ is given by $[X_1] \cdot [X_2] \equiv [X_1 \otimes X_2]$.

Formally, $U(V_{\mathrm{iso}(3,1)})$ represents the algebra of observables of our quantum relativistic system: the Hilbert space of states $\mathscr{H}$ of an elementary particle thus carries a representation of $U(V_{\mathrm{iso}(3,1)})$ describing all possible observations which we can make on the system. Such representation is, in fact, in a one-to-one correspondence with the representations of the Poincaré algebra itself. This is ensured by the so-called *universal property* of universal enveloping algebras (from which their name) which states that given any unital associative algebra U' for which there exist a linear map ϕ from a Lie algebra $\mathfrak{g}$ to U' which is a homomorphism, i.e., such that for $X_1, X_2 \in \mathfrak{g}$ one has that $\phi(X_1)\phi(X_2) - \phi(X_2)\phi(X_1) = \phi([X_1, X_2])$, there is a unique homomorphism of algebras between U' and the universal enveloping algebra $U(\mathfrak{g})$ which allows one to extend any linear map from $\mathfrak{g}$ to U' to a linear map from $U(\mathfrak{g})$ to U'. For more details about this property and its relevance for the one-to-one correspondence between the representations of $\mathfrak{g}$ and those of $U(\mathfrak{g})$, the reader can refer to [8].

As we described in detail above, in order to define observations on multi-particle states and antiparticle states, we need to introduce two additional operations on $U(V_{\mathrm{iso}(3,1)})$ (to make the notation lighter and since what we will say below is valid for any universal enveloping algebra, we will simply write U for $U(V_{\mathrm{iso}(3,1)})$ for the rest of this section), besides the product and the vector space structure already *built in*: the coproduct Δ and the antipode S. Such structures together with a series of compatibility conditions will turn U in what is known as *Hopf algebra*. Let us see where the compatibility conditions come from. First of all, let us notice that the unit element 1 we have introduced in the definition of the second quantized observable (5.3) and in the coproduct (5.4) *does not* belong to the Lie algebra. Indeed, the only product we have in the Lie algebra is the Lie bracket and the existence of a unit element such that $[1, X] = X$ for any element of the algebra would lead, through the Jacobi identity, to the non-sensical conclusion that $[X_1, X_2] = 0$ for *any* two elements of the Lie algebra. Of course, once we have constructed U we can define a unit element with respect to the associative product on U such that $1 \cdot X = X$ for any $X \in U$. Now the first condition that the coproduct must satisfy is the compatibility condition with the product on U namely that

$$\Delta(X_1 \cdot X_2) \equiv \Delta(X_1) \cdot \Delta(X_2),\tag{5.23}$$

i.e., that the coproduct is an algebra homomorphism of U. Such condition ensures that we can define proper representations of U on tensor product spaces. From this property, it immediately follows that the coproduct of the identity is idempotent and thus

$$\Delta(1) = 1 \otimes 1.\tag{5.24}$$

An element of an algebra with a coproduct of this type is said to be *group-like*. Notice that, turning things around, one could also say that the coproduct Δ equips U with a *co-algebra* structure and (5.23) is a condition which requires the product $\cdot$ to be a *co-algebra homomorphism*. In the spirit of this duality, we might expect that related to the associativity of the product $\cdot$, for which given $X_1, X_2, X_3 \in U$ one has that $(X_1 \cdot X_2) \cdot X_3 = X_1 \cdot (X_2 \cdot X_3)$, one should have a dual notion of *co-associativity*. Such condition is expressed by the following relation:

$$(\mathrm{id} \otimes \Delta) \circ \Delta = (\Delta \otimes \mathrm{id}) \circ \Delta,\tag{5.25}$$

where id is the identity operator on U. Such condition is telling us that when defining the action of an observable on states containing more than two particles, for example, in the definition (5.6), it does not actually matter if we use $\mathrm{id} \otimes \Delta$ or $\Delta \otimes \mathrm{id}$ to extend the action of the operator to states containing one more particle. Concretely, for the action of an observable on two tensor products of one-particle Hilbert spaces, we have

$$\Delta_3 \equiv (\mathrm{id} \otimes \Delta) \circ \Delta\mathcal{O} = \mathcal{O} \otimes \Delta 1 + 1 \otimes \Delta\mathcal{O} =$$
$$= \mathcal{O} \otimes 1 \otimes 1 + 1 \otimes \mathcal{O} \otimes 1 + 1 \otimes 1 \otimes \mathcal{O} = \Delta\mathcal{O} \otimes 1 + \Delta 1 \otimes \mathcal{O} = (\Delta \otimes \mathrm{id}) \circ \Delta\mathcal{O}.\tag{5.26}$$

The other characterizing feature of U as an algebra, besides the associativity, is the present of a unit element. On the co-algebra side, we can define a dual notion of *co-unit* . However, as with the coproduct, the co-unit will be introduced in terms of a map on U and thus we first need to define a unit *map* to which the co-unit will be dual. This unit map $e : \mathbb{C} \to U$ is such that

$$e(\lambda) = \lambda 1\tag{5.27}$$

foe every $\lambda \in \mathbb{C}$, where 1 is the unit element of U introduced above. The characterizing feature of the unit map is that the maps

$$e \otimes \mathrm{id} : \mathbb{C} \otimes U \longrightarrow U \otimes U\tag{5.28}$$

and

$$\mathrm{id} \otimes e : U \otimes \mathbb{C} \longrightarrow U \otimes U\tag{5.29}$$

when composed with the product reduce to the identity map on U

$$\cdot (e \otimes \mathrm{id}) = \mathrm{id} = \cdot(\mathrm{id} \otimes e) \tag{5.30}$$

under the isomorphism which relates $\mathbb{C} \otimes U$ and $U \otimes \mathbb{C}$ to U for which we can make the identifications

$$\lambda \otimes X \simeq \lambda X \simeq X \otimes \lambda. \tag{5.31}$$

In full (dual) analogy one defines a co-unit map $\varepsilon : U \to \mathbb{C}$ such that the composition of the coproduct with the maps

$$\varepsilon \otimes \mathrm{id} : U \otimes U \longrightarrow \mathbb{C} \otimes U \tag{5.32}$$

and

$$\mathrm{id} \otimes \varepsilon : U \otimes U \longrightarrow U \otimes \mathbb{C} \tag{5.33}$$

gives the identity map on U

$$(\varepsilon \otimes \mathrm{id}) \circ \Delta = \mathrm{id} = (\mathrm{id} \otimes \varepsilon) \circ \Delta. \tag{5.34}$$

In the same way we imposed the compatibility condition (5.23) between the product and the coproduct, we now impose that the co-unit is an algebra homomorphism, i.e.

$$\varepsilon(X_1 \cdot X_2) = \varepsilon(X_1)\varepsilon(X_2) \tag{5.35}$$

(the analogous condition for the unit map is trivially satisfied). In U the map satisfying these conditions is given by

$$\varepsilon(1) = 1, \qquad \varepsilon(X) = 0, \qquad \forall \, X \neq 1 \in U. \tag{5.36}$$

With the co-associative co-algebra structure associated to the coproduct and with the condition (5.35) on the co-unit the universal enveloping algebra U becomes a *bi-algebra* (to be distinguished from the *Lie* bi-algebra we encountered in the previous chapter).

At this point, we still have to find room to include in our construction the antipode map S which, as we saw, is naturally associated with the dual representation of a Lie algebra and thus of its universal enveloping algebra. It turns out that such map has a natural role in establishing a relationship between the unit/co-unit maps and the product/coproduct ones. In fact, let us notice that combining each of the two dual maps, we can define two endomorphisms of U

$$\cdot \circ \Delta : U \longrightarrow U \otimes U \longrightarrow U \tag{5.37}$$

and

$$\varepsilon \circ e : U \longrightarrow \mathbb{C} \longrightarrow U \tag{5.38}$$

which in principle have nothing to do with each other. However, we can connect such maps through the antipode by applying such map on one of terms of the coproduct as follows:

$$\cdot \circ (\mathrm{id} \otimes S) \circ \Delta = \varepsilon \circ e = \cdot \circ (S \otimes \mathrm{id}) \circ \Delta. \tag{5.39}$$

Let us see how this works explicitly. For our universal enveloping algebra U, the antipode is defined as

$$S(1) = 1, \qquad S(X) = -X, \qquad \forall\, X \neq 1 \in U. \tag{5.40}$$

The map $\varepsilon \circ e$ maps all elements of U into the zero element and the unit element into itself

$$\varepsilon \circ e(1) = 1 \qquad \varepsilon \circ e(X) = 0, \qquad \forall\, X \in U, \tag{5.41}$$

and ideed we have

$$\cdot \circ (\mathrm{id} \otimes S) \circ \Delta 1 = \cdot (1 \otimes 1) = 1 \tag{5.42}$$

for the unit element

$$\cdot \circ (\mathrm{id} \otimes S) \circ \Delta X = \cdot (X \otimes 1 - 1 \otimes X) = 0 = \cdot \circ (S \otimes \mathrm{id}) \circ \Delta X \tag{5.43}$$

for all the other elements of U. It can be proved [9] that the antipode is an algebra *anti-homomorphism*, i.e.

$$S(X_1 \cdot X_2) = S(X_2) \cdot S(X_1) \tag{5.44}$$

and the same property holds for the coproduct, i.e., S it is a co-algebra anti-homomorphism.

The unital associative algebra U equipped with a co-algebra structure, i.e., with a coproduct and co-unit map satisfying the properties (5.25) and (5.34) and an antipode for which the relation (5.39) holds is called a *Hopf algebra*.

We conclude this section by summarizing the structures introduced which, as we saw, arise naturally when considering quantum systems and their observables. In general, the latter are described by a Hopf algebra over the field of the complex numbers $\mathbb{C}$, that is a vector space $\mathcal{O}$ equipped with the following maps:

- The **product** $\cdot : \mathcal{O} \otimes \mathcal{O} \longrightarrow \mathcal{O}$

- The **unit** $e : \mathbb{C} \longrightarrow \mathcal{O}$

- The **coproduct** $\Delta : \mathcal{O} \longrightarrow \mathcal{O} \otimes \mathcal{O}$

- The **co-unit** $\varepsilon : \mathcal{O} \longrightarrow \mathbb{C}$

- The **antipode** $S : \mathcal{O} \longrightarrow \mathcal{O}$, such that the triple $(\mathcal{O}, \cdot, e)$ is a unital associative algebra, i.e.,, that the following properties hold:

- $\cdot \circ (\cdot \otimes \mathrm{id}) = \cdot \circ (\mathrm{id} \otimes \cdot)$ associativity

- $\cdot (e \otimes \mathrm{id}) = \mathrm{id} = \cdot (\mathrm{id} \otimes e)$ unit,

the triple $(\mathcal{O}, \Delta, \varepsilon)$ is a *co-unital co-associative* co-algebra, i.e., that one has

- $(\Delta \otimes \mathrm{id}) \circ \Delta = (\mathrm{id} \otimes \Delta) \circ \Delta$ co $-$ associativity

- $(\varepsilon \otimes \mathrm{id}) \circ \Delta = \mathrm{id} = (\mathrm{id} \otimes \varepsilon) \circ \Delta$ co $-$ unit,

and finally that the following relation between the five maps above holds

$$\cdot \circ (\mathrm{id} \otimes S) \circ \Delta = \varepsilon \circ e = \cdot \circ (S \otimes \mathrm{id}) \circ \Delta \quad \text{antipode axiom}.$$

The observables associated with the isometries if Minkowski spacetime are elements of the universal enveloping algebra of the Poincaré algebra $U(V_{\mathfrak{iso}(3,1)})$. These quantum observables are, in fact, elements of a Hopf algebra, given $X \in U(V_{\mathfrak{iso}(3,1)})$ we have the following co-algebra and antipode relations:

- **coproduct** $\Delta X = X \otimes 1 + 1 \otimes X$ and $\Delta 1 = 1 \otimes 1$

- **co-unit** $\varepsilon(1) = 1$, $\varepsilon(X) = 0$, $\forall X \neq 1 \in U$.

- **antipode** $S(1) = 1$, $S(X) = -X$, $\forall X \neq 1 \in U$.

We mentioned above that, strictly speaking, the unit element 1 (needed when defining the action of the generators on multi-partcle states) does not belong to the Poincaré Lie algebra $\mathfrak{iso}(3, 1)$. There is another object, which plays a fundamental role in quantum field theory, which also does not belong to $\mathfrak{iso}(3, 1)$ but is an element of the universal enveloping algebra $U(V_{\mathfrak{iso}(3,1)})$; the *Casimir operator*.

Casimir operators are defined in $U(V_{\mathfrak{iso}(3,1)})$ (and in any universal enveloping algebra) as a basis of its *center* (the set of elements which commute with all elements in $U(V_{\mathfrak{iso}(3,1)})$). For a generic universal enveloping algebra, such elements are of the form

$$\mathcal{C}_k = t^{i_1 \dots i_k} X_{i_1} \dots X_{i_k} \tag{5.45}$$

where $t^{i_1 \dots i_k}$ is a symmetric tensor invariant under the adjoint representation (see [8] for further details). The order k of this polynomial is called the order of the Casimir operator $\mathcal{C}_k$. Without going into detail, we recall the known fact that the Poincaré algebra has two Casimir operators, the quadratic *mass* Casimir

$$\mathcal{C}_2 = P_0^2 - \mathbf{P}^2 \tag{5.46}$$

and the fourth order Casimir

$$\mathcal{C}_4 = W_0^2 - \mathbf{W}^2 \tag{5.47}$$

obtained from the Pauli-Lubanski vector whose components are

$$W_0 = \mathbf{P} \cdot \mathbf{J}, \qquad \mathbf{W} = P_0 \mathbf{J} + \mathbf{P} \times \mathbf{K}, \tag{5.48}$$

where $\mathbf{J}$ and $\mathbf{K}$ are generators of rotations and boosts, respectively. Thanks to Schur's lemma (see [8] for details) we know that the Casimir operators are constant on irreducible representations of the Poincaré algebra, i.e., on one-particle states. Indeed, the eigenvalues of the quadratic Casimir give nothing but the *mass* squared of the particle

$$\mathcal{C}_2 |\mathbf{k}\rangle = m^2 |\mathbf{k}\rangle. \tag{5.49}$$

The physical significance of the Casimir $\mathcal{C}_4$ is immediate if we look at a particle at rest (since the eigenvalues of $\mathcal{C}_4$ are constant on states in an irreducible representation this leads to no loss of generality). Setting $\mathbf{P} = 0$ in (5.48) we have

$$W_0 = 0, \qquad \mathbf{W} = P_0 \mathbf{J}, \tag{5.50}$$

and since $P_0 = m$ on one-particle states, we have

$$\mathcal{C}_4 = -m^2 \mathbf{J}^2. \tag{5.51}$$

For particles with spin, the operator $\mathbf{J}^2$ characterizes their *spin*.[1] We thus obtain the well known fact that the quantum states representing elementary particles (irreducible representations of the Poincaré algebra) can be classified in term their Poincaré invariant notions of mass and spin.

Besides universal enveloping algebras, there is another important example of Hopf algebras which will play a fundamental role in what follows, these are the *algebras of functions on a group*. Let G be a group and $\mathscr{A}=\mathrm{Fun}(G)$ be the associative algebra

---

[1] For massless particles denoting their states with $|\mathbf{p}, h\rangle$ one has

$$W^\mu W_\mu |\mathbf{p}, h\rangle = W^\mu P_\mu |\mathbf{p}, h\rangle = P^\mu P_\mu |\mathbf{p}, h\rangle = 0. \tag{5.52}$$

Since W^μ is null on physical states and orthogonal to P^μ it must be proportional to the latter, i.e.

$$W^\mu |\mathbf{p}, h\rangle = h P^\mu |\mathbf{p}, h\rangle, \tag{5.53}$$

and the proportionality factor h is just the *helicity* of the particle with the helicity operator given by

$$\hat{h} = -\frac{\mathbf{P} \cdot \mathbf{J}}{|\mathbf{P}|}. \tag{5.54}$$

of complex functions on G with a unit element. The multiplication $(f_1, f_2) \to f_1 f_2$ and the unit I in $\mathscr{A}$ are defined by the formula

$$(f_1, f_2)(g) = f_1(g) f_2(g), \quad I(g) \equiv 1, \qquad g \in G. \tag{5.55}$$

The multiplication is a mapping $\mathscr{A} \times \mathscr{A} \to \mathscr{A}$ and the algebra $\mathscr{A}$ is commutative. The group operations allow us to introduce Hopf algebra mappings on $\mathscr{A}$, namely

- **coproduct** $\Delta : \mathscr{A} \equiv \text{Fun } G \to \text{Fun } (G \times G)$,
- **co-unit** $\varepsilon : \mathscr{A} \to \mathbb{C}$,
- **antipode** $S : \mathscr{A} \to \mathscr{A}$.

defined by the formulas

$$(\Delta f)(g_1, g_2) = f(g_1 g_2), \quad g_1 g_2 \in G,$$

$$\varepsilon(f) = f(1),$$

$$(Sf)(g) = f(g^{-1}), \quad g \in G. \tag{5.56}$$

It can be easily verified that these maps satisfy the axioms which make $\mathscr{A}$ a Hopf algebra. These structures will play a crucial role in the description of deformed translations of the κ-Poincaré algebra as we will see below.

Before concluding this section, we shall make one last important observation. As discussed at length in this section, universal enveloping algebras are naturally equipped with a Hopf algebra structure. One crucial feature of the co-algebra sector of these Hopf algebra is that the coproduct does not change if we flip its factors, namely

$$\sigma \circ \Delta X = \Delta X \tag{5.57}$$

where σ is the flip map

$$\sigma(X_1 \otimes X_2) = X_2 \otimes X_1. \tag{5.58}$$

A Hopf algebra for which such property holds is said to be *co-commutative*. This property is trivially satisfied by symmetry generators which act as derivatives on tensor products of representations, i.e., following the Leibniz rule. It turns out that there exist non-co-commutative generalizations of Hopf algebras, known as *quantum groups*, which admit non-trivial generalizations of the coproduct (5.4). The κ-Poincaré algebra which we will present in the following section is one example of such non-trivial Hopf algebras in which the non-co-commutative coproduct is strictly related to the non-abelian group manifold structure of momentum space.

5.2   The κ-Poincaré Hopf Algebra

We feel that in order to properly understand the new structures introduced by the κ-Poincaré algebra it is useful to first recall some basic facts about the structure of the standard Poincaré algebra.

Let us start from two very simple abelian Lie groups: the real numbers equipped with a group law given by standard addition, which we denote by $\mathbb{R}^+$ and the same set, excluding the zero element, with group law given by multiplication which we denote by $\mathbb{R}^\times$. We can combine these two groups using the *cartesian product* to build a larger group denoted by $\mathbb{R}^\times \times \mathbb{R}^+$ with multiplication law given by

$$(a_1, b_1)(a_2, b_2) = (a_1 a_2, b_1 + b_2), \qquad (a_1, b_1), \ (a_2, b_2) \in \mathbb{R}^\times \times \mathbb{R}^+, \quad (5.59)$$

i.e., by component-wise multiplication of the elements of the pairs in the cartesian product space. Notice how the group $\mathbb{R}^\times \times \mathbb{R}^+$ is abelian as its components are. The other notable feature of such group is that both the subgroups given by elements of the type $\{(a, 0)\} \simeq \mathbb{R}^\times$ and $\{(1, b)\} \simeq \mathbb{R}^+$ are *normal subgroups* of $\mathbb{R}^\times \times \mathbb{R}^+$, since they are invariant subgroups under the action by conjugation with an element $(c, d) \in \mathbb{R}^\times \times \mathbb{R}^+$

$$(c, d)(1, b)(c^{-1}, -d) = (1, d + b - d) = (b, 1) \tag{5.60}$$

$$(c, d)(a, 0)(c^{-1}, -d) = (cac^{-1}, 0) = (a, 0), \tag{5.61}$$

where we used the abelian character of ordinary addition and multiplication.

Now let us consider the group of transformations A of the real line $\mathbb{R}$ of the following form:

$$x \to ax + b, \qquad a \neq 0, \ b \in \mathbb{R}. \tag{5.62}$$

We can again denote elements of such group by ordered pairs (a, b) and see by applying two subsequent transformations that the group law is given by

$$(a_1, b_1)(a_2, b_2) = (a_1 a_2, a_1 b_2 + b_1), \tag{5.63}$$

and, as it can be easily verified, the group inversion is given by

$$(a, b)^{-1} = (a^{-1}, -a^{-1}b). \tag{5.64}$$

Notice how, again, the subgroups given by elements of the type $\{(a, 0)\}$ and $\{(1, b)\}$ are isomorphic to $\mathbb{R}^\times$ and $\mathbb{R}^+$, respectively, but now the group law (5.63) is *non-abelian*. Also, in this case, as it is easily verified, while $\mathbb{R}^+$ is still a normal subgroup of the group of transformations (5.62), the subgroup $\mathbb{R}^\times$ *is not*. Finally, the intersection of the two groups $\{(a, 0)\}$ and $\{(1, b)\}$ is the identity transformation $x \to x$, i.e., the element $(1, 0)$.

These features are what characterizes the so-called *semidirect product groups*. Precisely, let G be a group which splits into two subgroups H and N where N is a normal subgroup of H and $H \cap N = 1$ where 1 is the identity transformation. If every element of $g \in G$ can be written as a product $g = nh$ of some two elements $n \in N$ and $h \in H$, then G is said to be a semidirect product group and one writes $G = N \rtimes H$. The product of two elements of the group can be written as

$$g = g_1 g_2 = n_1 h_1 n_2 h_2 = n_1 h_1 n_2 h_1^{-1} h_1 h_2 \tag{5.65}$$

and since N is normal $h_1 n_2 h_1^{-1} \in N$ and so $n_1 h_1 n_2 h_1^{-1} \in N$. Thus $n_1 h_1 n_2 h_1^{-1} \in N$ and $h_1 h_2 \in H$ are, respectively, the first and second factor in the decomposition of the resulting group element $g = nh$. It is easy to see that in the case of the semidirect product group $A \equiv \mathbb{R}^\times \ltimes \mathbb{R}^+$ of transformations (5.62) the group law (5.63) corresponds to the general multiplication rule (5.65) indeed the conjugate element

$$(a_1, 0)(1, b_2)(a_1^{-1}, 0) = (a_1, 0)(a_1^{-1}, b_2) = (1, a_1 b_2) \tag{5.66}$$

and thus the composition of $(1, b_1)$ and $(1, a_1 b_2)$ gives the multiplication rule for translations found in (5.63).

Let us now look at the Lie algebra of the $ax + b$ group $A \equiv \mathbb{R}^\times \ltimes \mathbb{R}^+$. To do so, let us notice that $ax + b$ transformations are realized on functions of the real line by the following operators

$$T(b)\psi(x) = e^{-ibP} \psi(x) = \psi(x + b), \quad D(\alpha)\psi(x) = e^{-i\alpha R} \psi(x) = \psi(e^\alpha x), \tag{5.67}$$

where the generators of translations P and dilations R are given by

$$P = -i\frac{d}{dx}, \quad R = -ix\frac{d}{dx}. \tag{5.68}$$

These two operators satisfy the simplest non-abelian Lie algebra

$$[P, R] = -iP, \tag{5.69}$$

which, as we will soon see, in its four-dimensional generalization (i.e., adding two more abelian generators of translations) will be play a central role for κ-deformations. Let us now recall that given two Lie algebras $\mathfrak{g}_1$ and $\mathfrak{g}_2$ with Lie brackets $[\cdot, \cdot]_1$ and $[\cdot, \cdot]_2$, we can construct their direct sum Lie algebra $\mathfrak{g}_1 \oplus \mathfrak{g}_2$ which as a vector space is the direct sum of the $\mathfrak{g}_1$ and $\mathfrak{g}_2$ vector spaces equipped with the Lie bracket

$$[x, y] \equiv [x, y]_i \quad \text{for } x, y \in \mathfrak{g}_i, \quad i = 1, 2 \tag{5.70}$$

and

$$[x, y] = 0 \quad \text{if } x \in \mathfrak{g}_i, y \in \mathfrak{g}_j \ i \neq j. \tag{5.71}$$

When the last bracket is replaced by

$$[x, y] = z \quad \text{with } z \in \mathfrak{g}_1 \text{ when } x \in \mathfrak{g}_1 \ y \in \mathfrak{g}_2. \tag{5.72}$$

or, more concisely

$$[\mathfrak{g}_1, \mathfrak{g}_2] \subseteq \mathfrak{g}_1 , \tag{5.73}$$

the vector space $\mathfrak{g}_1 \oplus \mathfrak{g}_2$ is said to have the structure of a *semidirect product* Lie algebra. We thus see that the Lie algebra (5.69) is the simplest example of a semidirect sum of (abelian) Lie algebras.

The relevance of what we said so far for relativistic symmetries lies in the fact that the $ax + b$ group can be seen, loosely speaking, as a $0 + 1$ dimensional version of the Poincaré group. Indeed, as the $ax + b$ group, the Poincaré group $ISO(3, 1)$ is a semidirect product of the group of four-dimensional translations $\mathbb{R}^{3,1}$, now in Minkowski space, and the non-compact group of Lorentz transformations $SO(3, 1)$

$$ISO(3, 1) \equiv SO(3, 1) \ltimes \mathbb{R}^{3,1} . \tag{5.74}$$

A general Poincaré transformation on a Minkowski space four-vector x^μ is given by

$$x^\mu \rightarrow \Lambda^\mu{}_\nu x^\nu + b^\mu \quad , \tag{5.75}$$

which can be represented in terms of the action of a 5×5 matrix on a five-vector

$$\begin{pmatrix} x \\ 1 \end{pmatrix} \rightarrow \begin{pmatrix} \Lambda & b \\ 0 & 1 \end{pmatrix} \begin{pmatrix} x \\ 1 \end{pmatrix} \equiv (\Lambda, b) \begin{pmatrix} x \\ 1 \end{pmatrix} , \tag{5.76}$$

where we denoted with (Λ, b) a generic element of the Poincaré group $ISO(3, 1)$. Being a semidirect product group, we have that group multiplication is given by

$$(\Lambda_1, b_1)(\Lambda_2, b_2) = (\Lambda_1 \Lambda_2, \Lambda_1 b_2 + b_1) \tag{5.77}$$

with inverse given by

$$(\Lambda, b)^{-1} = (\Lambda^{-1}, -\Lambda^{-1} b) \tag{5.78}$$

Let us consider now a Poincaré transformation realized as a unitary operator $U(\Lambda, b)$ and let us write an infinitesimal transformation in terms of the generators of the Lorentz group $J^{\mu\nu}$ and of spacetime translations P^μ

$$U(1 + \omega, \varepsilon) = 1 + \frac{i}{2}\omega_{\mu\nu} J^{\mu\nu} - \varepsilon_\mu P^\mu , \tag{5.79}$$

where $\omega_{\mu\nu}$ and ε_μ are infinitesimal parameters (recall the the unit transformation is given by $(1, 0) \in ISO(3, 1)$) and from the invariance of the Minkowski metric it can be proved that $\omega_{\mu\nu}$ is antisymmetric. Let us now consider the action by conjugation of a generic element of $ISO(3, 1)$ on the infinitesimal element (5.79)

$$U(\Lambda, b) U(1 + \omega, \varepsilon) U^{-1}(\Lambda, b) = U(1 + \Lambda\omega\Lambda^{-1}, -\Lambda\varepsilon\Lambda^{-1}b + \Lambda\varepsilon), \tag{5.80}$$

in which we just used the group and inversion laws (5.77) and (5.78). Such expression once compared with

$$U(\Lambda, b)\, U(1+\omega, \varepsilon)\, U^{-1}(\Lambda, b) = U(\Lambda, b) \left(1 + \frac{i}{2}\omega_{\mu\nu} J^{\mu\nu} - \varepsilon_\mu P^\mu\right) U^{-1}(\Lambda, b)$$

(5.81)

leads to the following explicit expressions for the (adjoint) action of the group $ISO(3, 1)$ on its generators

$$U(\Lambda, b)\, P^\mu\, U^{-1}(\Lambda, b) = \Lambda_\nu{}^\mu P^{\,\nu}$$

(5.82)

$$U(\Lambda, b)\, J^{\mu\nu}\, U^{-1}(\Lambda, b) = \Lambda_\rho{}^\mu \Lambda_\sigma{}^\nu (J^{\rho\sigma} - b^\rho P^\sigma + b^\sigma P^\rho)\,.$$

(5.83)

In turn, these two relations can be used to derive the defining brackets of the Lie algebra $\mathfrak{iso}(3, 1)$. It suffices in fact to specialize (5.82) and (5.83) to the case

$$U(\Lambda, b) = U(1+\omega, \varepsilon) = 1 + \frac{i}{2}\omega_{\mu\nu} J^{\mu\nu} - \varepsilon_\mu P^\mu$$

(5.84)

with

$$U^{-1}(\Lambda, b) = U(1+\omega, \varepsilon) = 1 - \frac{i}{2}\omega_{\mu\nu} J^{\mu\nu} + \varepsilon_\mu P^\mu\,.$$

(5.85)

From (5.82), one obtains

$$[P^\mu, P^\nu] = 0$$

(5.86)

equating the terms in ε and

$$[P^\mu, J^{\rho\sigma}] = i\,(\eta^{\mu\sigma} P^\rho - \eta^{\mu\rho} P^\sigma)$$

(5.87)

from comparing the terms in ω. From (5.83), we obtain instead the commutators of the generators $J^{\mu\nu}$, i.e., the Lorentz algebra $\mathfrak{so}(3, 1)$

$$[J^{\mu\nu}, J^{\rho\sigma}] = i\,(\eta^{\mu\rho} J^{\nu\sigma} + \eta^{\sigma\mu} J^{\rho\nu} - \eta^{\nu\rho} J^{\mu\sigma} - \eta^{\sigma\nu} J^{\rho\mu})\,,$$

(5.88)

the $3 + 1$-dimensional analogue of the de Sitter algebra (3.8). The Lie brackets (5.86), (5.87) and (5.88) define the Poincaré Lie algebra, which can be expressed as the direct sum of the Lorentz algebra $\mathfrak{so}(3, 1)$ and the abelian subalgebra of translation generators $\mathbb{R}^{3,1}$ which is nothing but Minkowski space as a vector space, equipped with a trivial Lie bracket

$$\mathfrak{iso}(3, 1) = \mathfrak{so}(3, 1) \oplus \mathbb{R}^{3,1}$$

(5.89)

For later convenience, we write explicitly the generators of rotations and boosts as

$$J^i = \varepsilon^{ijk} J^{jk}\,, \qquad N^i = J^{i0}\,, \qquad i, j, k = 1, 2, 3\,,$$

(5.90)

(no summation over repeated indices in the first definition) in terms of which the commutators of the Poincaré algebra become

$$[J^i, J^j] = i\varepsilon^{ijk} J^k \tag{5.91}$$

$$[J^i, K^j] = i\varepsilon^{ijk} K^k \tag{5.92}$$

$$[K^i, K^j] = -i\varepsilon^{ijk} J^k \tag{5.93}$$

$$[J^i, P^j] = i\varepsilon^{ijk} P^k \qquad [J^i, P^0] = 0 \tag{5.94}$$

$$[K^i, P^j] = i\delta^{ij} P^0, \qquad [K^i, P^0] = i P^i . \tag{5.95}$$

Before moving on to the description of AN(3) group valued momenta as a starting point for introducing the κ-Poincaré alegebra, it will be useful to make a short digression on the meaning of 'positions' and 'momenta' when describing the phase space and symmetries of a relativistic particle. Let us denote with T the group of spacetime translation. For ordinary relativistic symmetries, this is just $\mathbb{R}^{3,1}$ seen as a group under addition. The Lie algebra t of translation generators, as a tangent space to the identity element, can be identified with $\mathbb{R}^{3,1}$ as vector spaces. The (trivial) Lie bracket on t is induced by the addition law of the group $T \equiv \mathbb{R}^{3,1}$. The dual group T^* is, by definition, given by equivalence classes of unitary irreducible representations of T and in the case $T \equiv \mathbb{R}^{3,1}$ elements of $T^* \equiv (\mathbb{R}^{3,1})^*$ are given by one-dimensional characters or in physics language 'plane waves'. When we write a plane wave like $e_p \in (\mathbb{R}^{3,1})^*$, we are simply saying that such element of the dual group $(\mathbb{R}^{3,1})^*$ has coordinates given by the four-vector p.

One usually refers to 'positions' as coordinates on Minkowski space, i.e., the translation group $T \equiv \mathbb{R}^{3,1}$. Indeed, in the usual description, the unreduced phase space of a spinless relativistic particle is given by the cotangent bundle of the group of translations T which is isomorphic [15] to $T \times $ t*. From this point of view, 'momenta' are just coordinates on the dual Lie algebra t*. Let us point out that one also speaks of 'momenta' when referring the spacetime translation generators, i.e., a basis of the Lie algebra t. In this case, spacetime 'coordinates' correspond to the basis of generators the dual algebra t*. In ordinary relativistic theories, we can refer to coordinates and momenta without specifying the objects we are referring to because T and t can be identified and so can their duals T^* and t*. When we say that the momentum becomes 'group valued' we mean that *the Lie algebra* t* *acquires non-trivial Lie brackets, i.e., it becomes non-abelian* (unlike the case of a particle in ordinary Minkowski space). This is to say that the dual group T^* is now a non-abelian group and thus momenta, as labels of plane waves, will obey a non-abelian composition rule. Let's first see what consequences this has, in general, and then we will specialize to the case of $AN(3)$ momenta.

First of all, according to the discussion in Chap. 4, a non-trivial Lie bracket on t* will correspond to a non-trivial Poisson-Lie structure on its dual algebra, i.e., coordinate functions x^μ on t will now have non-trivial Poisson brackets

$$[\cdot, \cdot]_{\text{t}^*} \neq 0 \longrightarrow \{\cdot, \cdot\}_{\text{t}} \neq 0 . \tag{5.96}$$

The second consequence is that a non-trivial Lie bracket on $\mathfrak{t}^*$ induces a non-trivial co-commutator $\delta : \mathfrak{t} \to \mathfrak{t} \otimes \mathfrak{t}$ on $\mathfrak{t}$ defined by

$$\delta(Y)(\xi_1, \xi_2) \equiv \langle Y, [\xi_1, \xi_2] \rangle , \qquad [\cdot, \cdot]_{\mathfrak{t}^*} \neq 0 \longrightarrow \delta(\cdot)_{\mathfrak{t}} \neq 0 . \tag{5.97}$$

Notice how even when the new structures are introduced the algebra of translation generators $\mathfrak{t}$ is still abelian and thus at the *Lie algebra level the generators of translations remain undeformed.*

We are now ready to describe how the AN(3) group valued momenta discussed in the previous chapters of this book can be included in a description of relativistic symmetries and how the new features introduced by such non-abelian group structure can be understood in the language of Hopf algebras. As the momentum space of a relativistic particle, the AN(3) group should have a well defined action of the Lorentz group SO(3, 1) on its elements. Such action can be obtained by noticing that both AN(3) and SO(3, 1) are subgroups of the de Sitter group SO(4, 1). This is evident from a splitting of the group SO(4, 1) known as *Iwasawa decomposition.* Such decomposition can be described starting from the Lie algebra $\mathfrak{so}(1, 4)$ which can be written as a direct sum of subalgebras

$$\mathfrak{so}(1, 4) = \hat{\mathfrak{t}} \oplus \hat{\mathfrak{n}} \oplus \hat{\mathfrak{a}} \tag{5.98}$$

where algebra $\hat{\mathfrak{t}}$ is the $\mathfrak{so}(1, 3)$ algebra, $\hat{\mathfrak{a}}$ is generated by the element

$$H = \begin{bmatrix} 0 & 0 & 0 & 0 & 1 \\ 0 & 0 & 0 & 0 & 0 \\ 0 & 0 & 0 & 0 & 0 \\ 0 & 0 & 0 & 0 & 0 \\ 1 & 0 & 0 & 0 & 0 \end{bmatrix} \tag{5.99}$$

and the algebra $\hat{\mathfrak{n}}$ by the elements

$$\mathfrak{n}_i = \begin{bmatrix} 0 & (\varepsilon_i)^T & 0 \\ \varepsilon_i & 0 & \varepsilon_i \\ 0 & -(\varepsilon_i)^T & 0 \end{bmatrix} \tag{5.100}$$

where ε_i are unit vectors in i'th direction ($\varepsilon_1 = (1, 0, 0)$, etc), and T denotes transposition. Every element of the algebra $\hat{\mathfrak{n}}$ is a negative root of the element H such that

$$[H, \hat{\mathfrak{n}}] = -\hat{\mathfrak{n}} . \tag{5.101}$$

Notice that this is nothing but the the Lie algebra (4.64) specialized to the case $n = 3$ under the identification $X^0 = -\frac{i}{\kappa} H$, $X^i = \frac{i}{\kappa} \mathfrak{n}_i$. From (5.98), it follows that every element g of the group SO(1, 4) can be decomposed as follows:

$$g = (\mathbf{k}\mathbf{n}\mathbf{a}) \quad \text{or} \quad g = (\mathbf{k}\,\vartheta\,\mathbf{n}\mathbf{a}) ,$$

Here $k \in SO(1, 3)$, the element a belongs to group A generated by H

$$A = \exp\left(i\, p_0 X^0\right) = \exp\left(\frac{p_0}{\kappa} H\right) = \begin{bmatrix} \cosh\frac{p_0}{\kappa} & 0 & 0 & 0 & \sinh\frac{p_0}{\kappa} \\ 0 & 1 & 0 & 0 & 0 \\ 0 & 0 & 1 & 0 & 0 \\ 0 & 0 & 0 & 1 & 0 \\ \sinh\frac{p_0}{\kappa} & 0 & 0 & 0 & \cosh\frac{p_0}{\kappa} \end{bmatrix} \tag{5.102}$$

and the element n belongs to group N generated by algebra $\hat{n} = p_i\, n_i$

$$N = \exp\left(i p_i X^i\right) = \exp\left(-\frac{1}{\kappa} p_i n_i\right) = \begin{bmatrix} 1 + \frac{1}{2\kappa^2}\mathbf{p}^2 & \frac{p_1}{\kappa} & \frac{p_2}{\kappa} & \frac{p_3}{\kappa} & \frac{1}{2\kappa^2}\mathbf{p}^2 \\ \frac{p_1}{\kappa} & 0 & 0 & 0 & \frac{p_1}{\kappa} \\ \frac{p_2}{\kappa} & 0 & 0 & 0 & \frac{p_2}{\kappa} \\ \frac{p_3}{\kappa} & 0 & 0 & 0 & \frac{p_3}{\kappa} \\ -\frac{1}{2\kappa^2}\mathbf{p}^2 & -\frac{p_1}{\kappa} & -\frac{p_2}{\kappa} & -\frac{p_3}{\kappa} & 1 - \frac{1}{2\kappa^2}\mathbf{p}^2 \end{bmatrix} \tag{5.103}$$

and $\vartheta = \text{diag}(-1,1,1,1,-1)$. Note that, as a result of isomorphism between the algebra $\hat{n} + \hat{a}$ (5.101) and the algebra (4.64), the element of NA can be equivalently expressed as a plane wave on κ-Minkowski spacetime with the time component appearing to the right (4.134)

$$\Pi(p) = \exp(i p_i X_i)\exp(i p_0 X^0). \tag{5.104}$$

Thus, we have that $\Pi(p)$ is an element of the non-abelian dual group $T^* = AN(3)$ while the κ-Minkowski coordinates X_μ are the generators of the Lie algebra $\mathfrak{t}^* = \mathfrak{an}_3$ dual to the Lie algebra of translation generators which remains abelian

$$[P_\mu, P_\nu] = 0 \tag{5.105}$$

and thus at the algebra level the translation sector remains unmodified. The non-abelian nature of $\mathfrak{t}^* = \mathfrak{an}_3$ will, however, affect the co-algebra structure of translation generators as we are going to show. The elements p_μ are coordinate functions on the group $AN(3)$ known as *bicrossporduct coordinates*. The quotient group structure of $AN(3) \simeq SO(1, 4)/SO(1, 3)$ allows us to obtain another system of coordinates which will be widely used in the coming chapters. Indeed, it is well known that the quotient of Lie groups $SO(1, 4)/SO(1, 3)$ is de Sitter space, a symmetric space with a transitive action of $SO(1, 4)$. Acting with the subgroup $AN(3)$ on the point $\mathscr{O}$

$$\mathscr{O} = \begin{bmatrix} 0 \\ 0 \\ 0 \\ 0 \\ \kappa \end{bmatrix} \tag{5.106}$$

we obtain the coordinate system (4.136)

$$P_0 = \kappa \sinh\frac{p_0}{\kappa} + \frac{\mathbf{p}^2}{2\kappa} e^{\frac{p_0}{\kappa}}$$

$$P_i = p_i e^{\frac{p_0}{\kappa}}$$

$$P_4 = \kappa \cosh \frac{p_0}{\kappa} - \frac{\mathbf{p}^2}{2\kappa} e^{\frac{p_0}{\kappa}} \tag{5.107}$$

on de Sitter space

$$-P_0^2 + P_i^2 + P_4^2 = \kappa^2 .$$

As discussed in Section (4.6), the coordinates p_0, p_i describe only half of de Sitter space, and therefore, the group KNA is not the whole group SO(1, 4). However, if we define

$$P_0 = \pm(\kappa \sinh \frac{p_0}{\kappa} + \frac{\mathbf{p}^2}{2\kappa} e^{\frac{p_0}{\kappa}})$$

$$P_i = \pm p_i e^{\frac{p_0}{\kappa}}$$

$$P_4 = \mp(\kappa \cosh \frac{p_0}{\kappa} - \frac{\mathbf{p}^2}{2\kappa} e^{\frac{p_0}{\kappa}}) \tag{5.108}$$

then the group SO(1,4) can be written in the form

$$\mathrm{SO}(1, 4) = \mathrm{KNA} \cup \mathrm{K}\vartheta\mathrm{NA} . \tag{5.109}$$

In order to derive the co-algebra structure of the the deformed translations, we will now use the fact that coordinate functions on AN(3) are a particular example of functions on a group, and thus have a natural Hopf algebra structure as we have seen in the previous section. The functions on the group that are going to be particularly useful in what follows are the coordinate functions on the AN(3) part of SO(4, 1)

$$f_\mu(g) = f_\mu (\exp(-ip_0 X_0) \exp(ip_i X_i)) = p_\mu = (p_0, p_i) \tag{5.110}$$

Looking at the group element (5.104) written as a plane wave and using the Baker–Campbell–Hausdorff formula

$$e^X e^Y = e^Z , \quad Z = X + Y + \frac{1}{2}[X, Y] + \frac{1}{12} ([X, [X, Y]] + [Y, [Y, X]) + \cdots \tag{5.111}$$

which in the particular case of $[X, Y] = sY$ takes the form

$$e^X e^Y = \exp \left(X + \frac{s}{1 - e^{-s}} Y \right) \tag{5.112}$$

it is easy to see that

$$\exp(-ip_0 X_0) \exp(ip_i X_i) = \exp(ip_i e^{-p_0/\kappa} X_i) \exp(-ip_0 X_0) \tag{5.113}$$

Using this property for the product of two group elements, we get

$$\exp(ip_i^{(1)}X_i)\exp(-ip_0^{(1)}X_0)\exp(ip_i^{(2)}X_i)\exp(-ip_0^{(2)}X_0)$$

$$\exp i\left(p_i^{(1)} + e^{-p_0^{(1)}/\kappa}p_i^{(2)}X_i\right)\exp -i(p_0^{(1)} + p_0^{(2)})X_0 \qquad (5.114)$$

while for the inverse, we have

$$(\exp(ip_iX_i)\exp(-ip_0X_0))^{-1} = \exp(-ip_iX_ie^{p_0/\kappa})\exp(ip_0X_0) \qquad (5.115)$$

Comparing these formulas with the definitions (5.56), we easily find that

$$\Delta(p_i) = p_i \otimes \mathbb{1} + e^{-p_0/\kappa} \otimes p_i , \qquad (5.116)$$

$$\Delta(p_0) = p_0 \otimes \mathbb{1} + \mathbb{1} \otimes p_0 , \qquad (5.117)$$

$$\varepsilon(p_0) = \varepsilon(p_i) = 0, \qquad (5.118)$$

$$S(p_0) = -p_0, \qquad S(p_i) = -p_ie^{p_0/\kappa} . \qquad (5.119)$$

Let us, for example, derive (5.116). The relevant formula for $f_i = p_i$ is

$$\Delta p_i(g_1, g_2) = p_i(g_1g_2) \qquad (5.120)$$

and examining (5.114) we get (5.116). Similarly, using

$$S(p)_i(g) = p_i(g^{-1}) \qquad (5.121)$$

and comparing with (5.115), we find (5.119). Thus, we see that at the algebra level κ-deformed translations are still abelian, while at the co-algebra level, they form a non-trivial Hopf algebra in which spatial translations have non-co-commutative coproduct and non-trivial antipode.

We now move on to the description of the Lorentz sector of the κ-Poincaré algebra. The first step is to observe that from the Iwasawa decomposition described above we can write an element of the group SO(1, 4) in two ways

$$K_1N_1A_1 = N_2A_2K_2 \qquad (5.122)$$

This decomposition is unique in the sense that once the left hand side is known all three elements on the right hand side are uniquely defined. The above equation can be also written as

$$K_1N_1A_1K_2^{-1} = N_2A_2 . \qquad (5.123)$$

Since this decomposition is unique the above equation defines the action of the Lorentz group $SO(1, 3)$ group on our deformed momenta. We can write explicitly

$$K_1 e^{i p_i x_i} e^{-i p_0 x_0} K_2^{-1} = e^{i p'_i x_i} e^{-i p'_0 x_0} \tag{5.124}$$

where $K_1, K_2 \in SO(1, 3)$. Let us solve this equation. To this end, we assume that K_2 depends on p_i, p_0 and K_1. Let us consider infinitesimal transformations, so that $K_1 - \mathbb{1} = \mathcal{K}_1$ and $K_2 - \mathbb{1} = \mathcal{K}_2$ are linear in the infinitesimal (boost) parameter ξ^i. Then we can use (5.124) to write an equation for the variation of the group element under a Lorentz transformation

$$\mathcal{K}_1 e^{i p_i x_i} e^{-i p_0 x_0} = \delta(e^{i p_i x_i} e^{-i p_0 x_0}) + e^{i p_i x_i} e^{-i p_0 x_0} \mathcal{K}_2 \tag{5.125}$$

where

$$\delta(e^{i p_i x_i} e^{-i p_0 x_0}) = e^{i p'_i x_i} e^{-i p'_0 x_0} - e^{i p_i x_i} e^{-i p_0 x_0}.$$

If we write $K_1 = \mathbb{1} + i \xi^i N_i$, where ξ^i are components of an infinitesimal boost parameter, we get

$$K_2 = \mathbb{1} + i \xi^i \left(e^{-p_0/\kappa} N_i + \frac{1}{\kappa} \varepsilon_{ijk} p_j M_k \right) \tag{5.126}$$

and

$$\delta(e^{i p_i x_i} e^{-i p_0 x_0})$$

$$= i \xi^k e^{i p_i x_i} \left(-p_k x_0 + \left(\delta_{kj} \left(\frac{\kappa}{2} \left(1 - e^{-2 p_0/\kappa} \right) + \frac{\mathbf{p}^2}{2\kappa} \right) - \frac{1}{\kappa} p_k p_j \right) x_j \right) e^{-i p_0 x_0}. \tag{5.127}$$

Using the above formulas, we can write Eq. (5.125) in the following form

$$(1 + i \xi^k N_k) e^{i p_i x_i} e^{-i p_0 x_0} \left(1 - i \xi^l \left(e^{-p_0/\kappa} N_l + \frac{1}{\kappa} \varepsilon_{ljk} p_j M_k \right) \right) \approx e^{i \mathbf{p'} \mathbf{x}} e^{-i p'_0 x_0} \tag{5.128}$$

which holds only up to linear order in ξ. Writing the leading order in ξ expansion of p'_i, p'_0 as

$$p'_j = p_j - i \xi^i [N_i, p_j], \quad p'_0 = p_0 - i \xi^i [N_i, p_0] \tag{5.129}$$

we find that the commutators are given by

$$[N_i, p_j] = i \delta_{ij} \left(\frac{\kappa}{2} \left(1 - e^{-2 p_0/\kappa} \right) + \frac{\mathbf{p}^2}{2\kappa} \right) - i \frac{1}{\kappa} p_i p_j, \quad [N_i, p_0] = i p_i. \tag{5.130}$$

These commutators describe the deformed action of boosts on group valued momenta for the particular choice of *bicrossproduct* coordinates (p_0, p_i) on the AN(3) momentum space. One of the characterizing features of these momentum variables is that the energy–momentum dispersion relation is non-linearly modified. This is, of course, strictly related to the nonlinear character of the boosts. The ordinary relativistic energy–momentum dispersion relation is, in fact, a sub-manifold, the *mass-shell*, of the $\mathbb{R}^{3,1}$ four-momentum space. In the κ-deformed context, the sub-manifolds of the AN(3) group which is left invariant by the action of the Lorentz group described above are the intersections of the $P_4 = const$ hyperplanes with the de Sitter hyperboloid [10]. Such deformed mass-shells are given by

$$\mathscr{C}(p) = \left(2\kappa \sinh \frac{p_0}{s\kappa}\right)^2 - \mathbf{p}^2 e^{\frac{p_0}{\kappa}} . \tag{5.131}$$

When expanded in powers of $1/\kappa$, this modified dispersion relation provides an example of Planck scale deformation of the mass-shell which have been widely studied as candidate quantum gravity effects on particle kinematics (see, e.g., [11, 12] and references therein), and to explore the trans-planckian regime of semiclassical gravity effects like Hawking radiation [13] and early universe cosmology [14].

We are now ready to derive the co-algebra operations on the κ-Poincaré deformed boost generators. Using the notation introduced above, we can write the infinitesimal action of the boost on functions on the group $g \in$ AN(3) as

$$(\mathbb{1} + i\xi^i N_i).f(g) = f\left((\mathbb{1} + i\xi^i N_i)gK_2^{-1}\right) \tag{5.132}$$

where dot means action by commutator.

The antipode $S(N_i)$ can be defined from the action of infinitesimal Lorentz boosts on inverse elements as follows:

$$(\mathbb{1} + i\xi^i N_i)(e^{ip_i x_i} e^{-ip_0 x_0})^{-1} = e^{ip'_i x_i} e^{-ip'_0 x_0}\left(\mathbb{1} - i\xi^i S(N_i)\right) \tag{5.133}$$

Recalling expression (5.115), we see that all we have to do to find the antipode is to use (5.126, where instead of p_μ we insert $S(p_\mu)$ and then change the overall sign). As a result, we find easily the explicit form of the antipode $S(N_i)$

$$S(N_i) = -e^{\frac{p_0}{\kappa}}\left(N_i - \frac{1}{\kappa}\varepsilon_{ijk}p_j M_k\right) . \tag{5.134}$$

Deriving the coproduct is a bit more complicated. Let us first rewrite the definition of coproduct in a more convenient form

$$(\Delta f)(g_1, g_2) = \sum_\alpha f_\alpha^{(1)} \otimes f_\alpha^{(2)}(g_1, g_2) = f(g_1 g_2). \tag{5.135}$$

where α is a label and the above formula must be understood as multiplication

$$\sum_\alpha f_\alpha^{(1)} \otimes f_\alpha^{(2)}(g_1, g_2) = \sum_\alpha f_\alpha^{(1)}(g_1) f_\alpha^{(2)}(g_2) .$$

Using this abstract formula, let us derive the coproduct of N_i from their action on products of elements belonging to group AN(3). This means that we can act either on the product of of two AN(3)) group elements or first on the elements and then take the product. We have

$$f\left(\mathsf{K}_1 g_1 g_2 \mathsf{K}_2^{-1}\right) = f\left(\mathsf{K}_1 g_1 h^{-1} h g_2 \mathsf{K}_2^{-1}\right) = \sum_\alpha f_\alpha^{(1)} \otimes f_\alpha^{(2)} (\mathsf{K}_1 g_1 h^{-1}, h g_2 \mathsf{K}_2^{-1}),$$
(5.136)

for $\mathsf{K} = \mathbb{1} + i\xi^i N_i$ we can again use (5.126) to find an expression for h and then for K_2 and then to finally obtain the following coproduct:

$$\Delta(N_i) = N_i \otimes 1 + e^{-p_0/\kappa} \otimes N_i + \frac{1}{\kappa} \varepsilon_{ijk} p_j \otimes M_k.$$
(5.137)

Finally, for the co-unit, we have

$$\varepsilon(N_i).f(g) = N_i.(\varepsilon f)(g) \implies \varepsilon(N_i) = 0.$$
(5.138)

We have thus derived all the co-algebra maps for the boost generators. One can proceed in a similar fashion for the generators of rotations M_i and discover that both at algebra and co-algebra level they exhibit undeformed commutators and have trivial co-algebra maps. Below, we summarize the deformed algebra and co-algebra sectors

$$[M_i, M_j] = i\, \varepsilon_{ijk} M_k, \quad [M_i, N_j] = i\, \varepsilon_{ijk} N_k,$$
(5.139)

$$[N_i, N_j] = -i\, \varepsilon_{ijk} M_k.$$
(5.140)

$$[M_i, p_j] = i\, \varepsilon_{ijk} p_k, \quad [M_i, p_0] = 0$$
(5.141)

$$[N_i, p_j] = i\, \delta_{ij} \left(\frac{\kappa}{2}\left(1 - e^{-2p_0/\kappa}\right) + \frac{\mathbf{p}^2}{2\kappa}\right) - i\frac{1}{\kappa} p_i p_j, \quad [N_i, p_0] = i\, p_i,$$
(5.142)

and

$$\Delta(p_i) = p_i \otimes \mathbb{1} + e^{-p_0/\kappa} \otimes p_i,$$
(5.143)

$$\Delta(p_0) = p_0 \otimes \mathbb{1} + \mathbb{1} \otimes p_0,$$
(5.144)

$$S(p_0) = -p_0, \quad S(p_i) = -p_i e^{p_0/\kappa}.$$
(5.145)

$$\varepsilon(p_0) = \varepsilon(p_i) = 0,$$
(5.146)

$$\Delta(N_i) = N_i \otimes 1 + e^{-p_0/\kappa} \otimes N_i + \frac{1}{\kappa} \varepsilon_{ijk} p_j \otimes M_k.$$
(5.147)

$$S(N_i) = -e^{\frac{p_0}{\kappa}}\left(N_i - \frac{1}{\kappa} \varepsilon_{ijk} p_j M_k\right), \quad \varepsilon(N_i) = 0$$
(5.148)

$$\Delta(M_i) = M_i \otimes 1 + 1 \otimes M_i, \quad S(M_i) = -M_i, \quad \varepsilon(M_i) = 0. \tag{5.149}$$

These relations characterize the κ-Poincaré Hopf algebra in the so-called bicrossprod-uct basis. As we will see in the next chapters, other coordinate systems on the AN(3) group manifold will exhibit different algebra and co-algebra relations. In particular, the embedding coordinates (4.136) will be associated to a variant of the κ-Poincaré algebra in which the algebra sector is *undeformed* and all the non-trivial structures appear only in the co-algebra sector.

In the remaining chapters of these notes, we will sketch the construction of clas-sical and quantum fields based on these κ-deformed symmetries and highlight their most salient features.

References

1. Reed, M., Simon, B.: Methods of Modern Mathematical Physics. 1. Functional Analysis. Aca-demic Press, New York (1972)
2. Horodecki, R., Horodecki, P., Horodecki, M., Horodecki, K.: Rev. Mod. Phys. **81**, 865–942 (2009). arXiv:quant-ph/0702225 [quant-ph]
3. Cook, J.M.: The mathematics of second quantization. Proc. Nat. Acad. Sci. U. S. A. **37**, 417–420 (1951). ISSN 0027-8424
4. Humphreys, J.E.: Introduction to Lie Algebras and Representation Theory, 171p. Springer, New York, USA (1980)
5. Wald, R.M.: Quantum Field Theory in Curved Space-Time and Black Hole Thermodynamics, 205p. University Press, Chicago, USA (1994)
6. Geroch, R.P.: Quantum Field Theory: 1971 Lecture Notes. Minkowski Institute Press, Montreal (2013)
7. Geroch, R.P.: Mathematical Physics. The University of Chicago Press (1985)
8. Fuchs, J., Schweigert, C.: Symmetries, Lie algebras and representations: a graduate course for physicists. Cambridge Monographs on Mathematical Physics (2003)
9. Majid, S.: Foundations of Quantum Group Theory. Cambridge University Press (1996)
10. Kowalski-Glikman, J., Nowak, S.: Doubly special relativity and de Sitter space. Class. Quant. Grav. **20**, 4799–4816 (2003). arXiv:hep-th/0304101 [hep-th]
11. Amelino-Camelia, G.: Quantum-spacetime phenomenology. Living Rev. Rel. **16**, 5 (2013). arXiv:0806.0339 [gr-qc]
12. Hossenfelder, S.: Minimal length scale scenarios for quantum gravity. Living Rev. Rel. **16**, 2 (2013). arXiv:1203.6191 [gr-qc]
13. Corley, S., Jacobson, T.: Hawking spectrum and high frequency dispersion. Phys. Rev. D **54**, 1568–1586 (1996). arXiv:hep-th/9601073 [hep-th]
14. Brandenberger, R.H.: Lectures on the theory of cosmological perturbations. Lect. Notes Phys. **646**, 127–167 (2004). arXiv:hep-th/0306071 [hep-th]
15. Abraham, R., Marsden, J.E.: Foundation of Mechanics. Benjamin/Cummings Publishing Com-pany (1978)

Classical Fields, Symmetries, and Conserved Charges

We will devote the two final chapters of the book to present some elements of classical and quantum field theory on non-commutative κ-Minkowski space, for which the κ-Poincaré algebra is an algebra of symmetries. We base our presentation below on the papers [1–3].

There are many approaches to field theory on κ-Minkowski space [4–16] and the reader may be willing to consult these papers to see the differences in motivations and results. The reader interested in field theory can go directly to Sect. 6.1.4, where the brief description off all technical tools required for the construction of field theory is summarized.

6.1   Preliminaries

In this section, we will start the construction of κ-deformed field theory with the description of mathematical tools that we will have to make use of later. To make this exposition self-contained, we will recall here some notions that appeared already in the preceding chapters. In this chapter, we will use a slightly different notation than we did in the previous one: we denote the coordinates on non-commutative κ-Minkowski space as $\hat{x}^\mu$ and on the commutative Minkowski space as x^μ; momenta in the bicrossproduct basis will be denoted by $k, l, \ldots$ and in the classical one $p, q, \ldots$.

We start with recalling once again the defining commutators of κ-Minkowski space [17]

$$[\hat{x}^0, \hat{x}^i] = \frac{i}{\kappa}\hat{x}^i, \qquad [\hat{x}^i, \hat{x}^j] = 0, \qquad i, j = 1, 2, 3. \tag{6.1}$$

As we saw in the previous chapters, these commutators close the Lie algebra $\mathsf{an}(3)$ which can be seen as the product of an abelian algebra A generated by $\hat{x}^0$ and the one

© Springer-Verlag GmbH Germany, part of Springer Nature 2021
M. Arzano and J. Kowalski-Glikman, *Deformations of Spacetime Symmetries*,
Lecture Notes in Physics 986,
https://doi.org/10.1007/978-3-662-63097-6_6

generated by three nilpotet generators $\hat{x}^i$. As before, we consider the five-dimensional matrix representation (4.133) of the generators in (6.1)

$$\hat{x}^0 = -\frac{i}{\kappa}\begin{pmatrix} 0 & 0 & 1 \\ 0 & 0 & 0 \\ 1 & 0 & 0 \end{pmatrix} \quad \hat{x} = \frac{i}{\kappa}\begin{pmatrix} 0 & \varepsilon^T & 0 \\ \varepsilon & 0 & \varepsilon \\ 0 & -\varepsilon^T & 0 \end{pmatrix} . \tag{6.2}$$

By exponentiating the generators of the algebra (6.1), we obtain the group element (non-commutative plane wave)

$$\hat{e}_p = e^{ip_i\hat{x}^i}e^{ip_0\hat{x}^0} , \tag{6.3}$$

which can be written in schematic form

$$\hat{e}_k = \begin{pmatrix} \frac{\bar{p}_4}{\kappa} & \frac{\mathbf{p}}{\kappa} & \frac{p_0}{\kappa} \\ \frac{\mathbf{p}}{\kappa} & 1 & \frac{\mathbf{p}}{\kappa} \\ \frac{\bar{p}_0}{\kappa} & -\frac{\mathbf{p}}{\kappa} & \frac{p_4}{\kappa} \end{pmatrix} , \tag{6.4}$$

where p_0, p_i, and p_4 will be defined below, while $\bar{p}_0 = \kappa \sinh\frac{k_0}{\kappa} - \frac{\mathbf{k}^2}{2\kappa}$, $\bar{p}_4 = \kappa \cosh\frac{k_0}{\kappa} + \frac{\mathbf{k}^2}{2\kappa}e^{k_0/\kappa}$.

As we saw in the previous chapter, the group manifold of AN(3) can be constructed by choosing a point in five-dimensional Minkowski space, which becomes the momentum space origin $\mathscr{O}$ with coordinates $(0, \ldots, 0, \kappa)$ and acting on it with the matrix $\hat{e}_k$ (6.4), obtaining

$$(p_0, p_i, p_4) = \hat{e}_k \mathscr{O}$$

On the left hand side, we have coordinates of a point in the five-dimensional Minkowski space, being in one-to-one correspondence with the group element $\hat{e}_k$. The coordinates (p_0, p_i, p_4) are related to the original parametrization (k_0, k_i) of the group element as follows:

$$p_0(k_0, \mathbf{k}) = \kappa \sinh\frac{k_0}{\kappa} + \frac{\mathbf{k}^2}{2\kappa}e^{k_0/\kappa},$$
$$p_i(k_0, \mathbf{k}) = p_i\, e^{k_0/\kappa},$$
$$p_4(k_0, \mathbf{k}) = \kappa \cosh\frac{k_0}{\kappa} - \frac{\mathbf{k}^2}{2\kappa}e^{k_0/\kappa}. \tag{6.5}$$

These coordinates satisfy the equation[1]

$$-p_0^2 + \mathbf{p}^2 + p_4^2 = \kappa^2, \quad p_4 > 0 \tag{6.6}$$

---

[1] There are two solutions of the first equation in (4.137), but since the point $\mathscr{O}$ for which $p_4 = 1$ must belong to the relevant solution we choose p_4 positive.

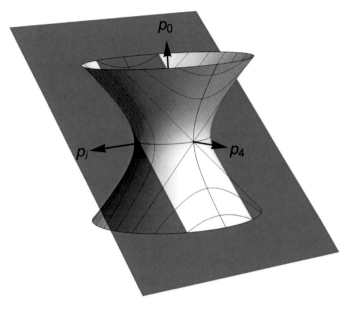

Fig. 6.1 AN(3) momentum space is a submanifold of de Sitter space subject to the conditions $p_0 + p_4 > 0$, $p_4 > 0$ and is depicted as a part of the manifold above the plane. The vertical hyperbolas are orbits of Lorentz group

and thus the group AN(3) is isomorphic, as a manifold, to the submanifold of the 4-dimensional de Sitter space (see Fig. 6.1) defined by

$$p_0 + p_4 = e^{k_0/\kappa} > 0, \quad p_4 \equiv \sqrt{\kappa^2 + p_0^2 - \mathbf{p}^2} > 0. \tag{6.7}$$

The coordinates p_μ on AN(3) are called embedding coordinates.

The subgroup of the group of isometries of five-dimensional Minkowski space, SO(4, 1) that preserves the point $\mathscr{O}$ is isomorphic to Lorentz group SO(3, 1)

$$\begin{pmatrix} \text{SO(3, 1)} & \mathbf{0} \\ \mathbf{0} & 1 \end{pmatrix} \begin{pmatrix} \mathbf{0} \\ \kappa \end{pmatrix} = \begin{pmatrix} \mathbf{0} \\ \kappa \end{pmatrix}$$

Therefore, under action of this subgroup, that we will identify with Lorentz symmetry the components p_μ transform as components of a Lorentz vector, while p_4 is a Lorentz scalar. Infinitesimally, for rotation δ_ρ and boost δ_λ, we have

$$\delta_\rho p_0 = 0, \quad \delta_\rho p_i = \varepsilon_{jik}\, \rho_j\, p_k$$
$$\delta_\lambda p_0 = \lambda_i\, p_i, \quad \delta_\lambda p_i = \lambda_i\, p_0$$

In what follows, we will, exchangeably, denote these relations with the help of the action of rotation and boost generators M_i and N_i

$$M_i \triangleright p_0 = 0, \quad M_i \triangleright p_j = \varepsilon_{ijk}\, p_k$$
$$N_i \triangleright p_0 = p_i, \quad N_i \triangleright p_j = \delta_{ij}\, p_0 \tag{6.8}$$

As a result of this simple standard Lorentz action, $p^2 = p_\mu p^\mu$ is a Lorentz scalar, and as usual, we can make use of it to define the mass-shell condition

$$p_0^2 - \mathbf{p}^2 = m^2 \tag{6.9}$$

Notice that on shell, the scalar p_4 takes the particularly simple form

$$p_4 = \sqrt{m^2 + \kappa^2}. \tag{6.10}$$

The condition $p_0 + p_4 > 0$ takes a particularly simple form of shell

$$p_0 + \sqrt{m^2 + \kappa^2} > 0. \tag{6.11}$$

One can observe that this condition is identically satisfied for the positive energy states $p_0 > 0$, but restricts the range of allowed negative energies. Further, it looks, naively, manifestly Lorentz violating: by Lorentz transformation we can transform any negative p_0 to an arbitrarily large negative value. We will resolve this problem in the next subsection.

It turns out that in order to construct field theory, we have to introduce another, antipodal momentum space defined as an orbit of AN(3) group emanating from the point $\mathcal{O}^*$ with coordinates $(0, \ldots, 0, -\kappa)$. The coordinates on this space can be constructed with the help of a special element ϑ [1] that maps $(0, \ldots, 0, \kappa)$ to $(0, \ldots, 0, -\kappa)$,

$$\vartheta = e^{\pi \kappa \hat{x}^0} = \begin{pmatrix} -1 & \mathbf{0} & 0 \\ \mathbf{0} & \mathbf{1} & \mathbf{0} \\ 0 & \mathbf{0} & -1 \end{pmatrix} \tag{6.12}$$

(or $\hat{e}_k$ in (6.4) with $k_i = 0$, $k_0 = -i\pi\kappa$).

We define

$$\hat{e}_k^* = \hat{e}_k \vartheta = e^{ik_i \hat{x}^i} e^{ik_0 \hat{x}^0} \vartheta \tag{6.13}$$

and acting with this group element on $(0, \ldots, 0, \kappa)$, instead of (6.5) we get

$$p_0^*(k_0, \mathbf{k}) = -\kappa \sinh \frac{k_0}{\kappa} - \frac{\mathbf{k}^2}{2\kappa} e^{k_0/\kappa},$$

$$p_i^*(k_0, \mathbf{k}) = -p_i \, e^{k_0/\kappa},$$

$$p_4^*(k_0, \mathbf{k}) = -\kappa \cosh \frac{k_0}{\kappa} + \frac{\mathbf{k}^2}{2\kappa} e^{k_0/\kappa}. \tag{6.14}$$

with

$$p_0^* + p_4^* = -e^{k_0/\kappa} < 0, \quad p_4^* \equiv -\sqrt{\kappa^2 + (p_0^*)^2 - (\mathbf{p}^*)^2} < 0. \tag{6.15}$$

6.1.1 More on the κ-Poincaré Algebra

Let us now go back to the κ-Poincaré algebra which we introduced in the last chapter. One point of view for understanding the deformation which characterizes such algebra is to link the non-commutativity of spacetime to momenta which belong to a non-abelian Lie group rather than to a vector space as in ordinary relativistic kinematics. To this end, let us consider the product of two 'non-commutative plane waves' $\hat{e}_k = e^{ik_i\hat{x}^i}e^{ik_0\hat{x}^0}$ and $\hat{e}_l = e^{il_i\hat{x}^i}e^{il_0\hat{x}^0}$. This can be written as

$$\hat{e}_k\hat{e}_l = e^{ip_i\hat{x}^i}e^{ip_0\hat{x}^0}e^{il_i\hat{x}^i}e^{il_0\hat{x}^0} = e^{i(p_i\oplus l_i)\hat{x}^i}e^{i(p_0\oplus l_0)\hat{x}^0} \equiv \hat{e}_{k\oplus l}, \tag{6.16}$$

where

$$k_i \oplus l_i = k_i + e^{-k_0/\kappa}l_i, \quad k_0 \oplus l_0 = k_0 + l_0 \tag{6.17}$$

The composition law $k_\mu \oplus l_\mu = (k_0 \oplus l_0, k_i \oplus l_i)$ is clearly non-abelian, i.e., $k_\mu \oplus l_\mu \neq l_\mu \oplus k_\mu$, since the AN(3) group is non-abelian. As discussed in detail in Chap. 5, the composition law for momenta reflects the composition of conserved quantum numbers associated to translation generators. In particular, the familiar addition of momenta can be seen as a consequence of the Leibiniz rule for the action of translation generators on multi-particle states. The non-abelian composition of momenta thus reflects a deformed action of space translation generators. As we know from the preceding chapter, in the language of Hopf algebras this can be expressed in terms of a non-trivial coproduct for the spatial translation generators that we also denote k_i

$$\Delta k_i = k_i \otimes \mathbb{1} + e^{-k_0/\kappa} \otimes k_i, \tag{6.18}$$

while the time translation generator acts according to the usual Leibniz rule expressed by the trivial coproduct

$$\Delta k_0 = k_0 \otimes \mathbb{1} + \mathbb{1} \otimes k_0. \tag{6.19}$$

In a similar fashion, the group inversion is reflected in a non-trivial antipode for the generators

$$S(k_0) = -k_0, \quad S(k_i) = -e^{k_0/\kappa}k_i, \tag{6.20}$$

which determines the appropriate generalization of momentum subtraction operation $\ominus$ needed in order for the basic relation $k_\mu \oplus (\ominus k_\mu) = 0$ to hold. We provided a derivation of the coproduct and antipode for the remaining κ-Poincaré generators read in the last chapter and they are given by

$$\Delta M_i = M_i \otimes \mathbb{1} + \mathbb{1} \otimes M_i, \quad S(M_i) = -M_i,$$

$$\Delta N_i = N_i \otimes \mathbb{1} + e^{-k_0/\kappa} \otimes N_i + \frac{1}{\kappa}\varepsilon_{ijk}k^j \otimes M^k,$$

$$S(N_i) = -e^{k_0/\kappa}N_i + \frac{1}{\kappa}\varepsilon_{ijk}e^{k_0/\kappa}k^j M^k. \tag{6.21}$$

This particular realization of the κ-Poincaré Hopf algebra in terms of the generators $\{k_\mu, M_i, N_i\}$, i.e., is known as the bicrossproduct basis of the κ-Poincaré algebra [17]. One of the characterizing features of this basis is that the commutators between boosts and translation generators are deformed

$$[k_0, N_i] = -ik_i ,$$

$$[k_i, N_j] = -i\delta_{ij} \left(\frac{\kappa}{2} \left(1 - e^{-2k_0/\kappa} \right) + \frac{1}{2\kappa} k_i k^i \right) + \frac{i}{\kappa} k_i k_j . \tag{6.22}$$

and the invariant Casimir in this basis reads

$$\mathscr{C}(k) = \left(2\kappa \sinh \left(\frac{k_0}{2\kappa} \right) \right)^2 - \mathbf{k}^2 e^{k_0/\kappa} \tag{6.23}$$

As we anticipated in Chap. 5, different choices of coordinates on the AN(3) manifold will lead, in general, to different coproducts and antipodes for the associated translation generators. These new expressions for the coproducts and antipodes are obtained from the old one by combining the old ones with the mapping that transforms one coordinates into another. For example, the relations above can be used to derive the coproducts and antipodes for translation generators p_μ associated to the embedding coordinates defined in (6.5)

$$\Delta(p_0) = p_0 \otimes p_+ + p_+^{-1} \otimes p_0 + \frac{1}{\kappa} \sum_{i=1}^{3} p_i p_+^{-1} \otimes p_i ,$$

$$\Delta(p_i) = p_i \otimes p_+ + 1 \otimes p_i ,$$

$$\Delta(p_4) = p_4 \otimes p_+ - p_+^{-1} \otimes p_0 - \frac{1}{\kappa} \sum_{i=1}^{3} p_i p_+^{-1} \otimes p_i \tag{6.24}$$

$$S(p_0) = -p_0 + \frac{1}{\kappa} \mathbf{p}^2 p_+^{-1} = \kappa p_+^{-1} - p_4 , \qquad S(p_i) = -p_i p_+^{-1} , \qquad S(p_4) = p_4 , \tag{6.25}$$

where $p_+ \equiv (p_0 + p_4)/\kappa$. These generators are known in the literature as the classical basis of the κ-Poincaré algebra since, unlike the bicrossproduct basis reviewed above, their commutators are just the ones of the ordinary Poincaré algebra.

$$[p_0, N_i] = -ip_i ,$$

$$[p_i, N_j] = -i\delta_{ij} p_0 . \tag{6.26}$$

Therefore, the standard Casimir is Lorentz invariant and we can make use os the standard mass-shell relation

$$\mathscr{C} = p_0^2 - \mathbf{p}^2 = m^2 \tag{6.27}$$

The corresponding coproducts and antipodes for rotations and boosts will be now given by

$$\Delta(M_i) = M_i \otimes 1 + 1 \otimes M_i$$

$$\Delta(N_i) = N_i \otimes 1 + p_+^{-1} \otimes N_i + \varepsilon_{ijk} \frac{1}{\kappa} p_j p_+^{-1} \otimes M_k \qquad (6.28)$$

and

$$S(M_i) = -M_i$$

$$S(N_i) = -N_i p_+ + \varepsilon_{ijk} \frac{1}{\kappa} p_j M_k . \qquad (6.29)$$

Let us now return to the problem of apparent Lorentz non-invariance of the condition $p_0 + p_4 > 0$ defining the AN(3) group manifold as a submanifold of de Sitter space. Let us start recalling the boost along the first axis, with the boost parameter ξ

$$p_0' = p_0 \cosh \xi + p_1 \sinh \xi , \quad p_1' = p_1 \cosh \xi + p_0 \sinh \xi \qquad (6.30)$$

Clearly, since for positive energy particle $p_0 > |p_1|$, if $p_0 > 0$ then $p_0' > 0$, and therefore, $p_0 + p_4 > 0$. However, if $p_0 < 0$ p_0' can take an arbitrary negative value and $p_0' + p_4$ can be negative [19]. This can be seen in the Fig. 6.1, where for negative energies the Lorentz boost orbits pierce through the plane and go outside the AN(3) manifold.

To solve this problem, let us observe that for any positive $p_0 > 0$ its antipode satisfies

$$S(p_0) = -p_0 + \frac{\mathbf{p}^2}{p_0 + p_4} = \frac{\kappa^2}{p_0 + p_4} - p_4$$

so that

$$S(p_0) + p_4 > 0$$

and this expression goes to zero for $p_0 \to \infty$. We conclude, therefore, that the antipode is an one-to-one map between the positive energy submanifold of the AN(3) momentum space and the negative energy one. We can now define an action of the Lorentz group on the momentum space as follows:

$$(\Lambda \triangleright p)_\mu = \Lambda_\mu{}^\nu p_\nu , \quad \text{for } p_0 > 0$$

$$(\Lambda \triangleright p)_\mu = S(\Lambda \triangleright S(p))_\mu , \quad \text{for } p_0 < 0, \ p_0 + p_4 > 0 \qquad (6.31)$$

(For negative energy p, $S(p)$ is a positive energy one; its Lorentz transformation is positive energy again and finally its antipode is negative energy again.)

It will be instructive to see how the antipode transforms under Lorentz boosts ($S(p)_0$ is a rotation scalar, while the components of $S(p)_i$ form a vector under rotation.)

Let us investigate properties of this transformation in the case of a infinitesimal Lorentz boost with parameter λ^i

$$\delta_\lambda p_i = \lambda_i \, p_0 \,, \quad \delta_\lambda p_0 = \lambda^i \, p_i \tag{6.32}$$

Remembering that p_4 is Lorentz invariant using (6.25), we find

$$\delta_\lambda S(p_0) = -\lambda^i \, p_i + \frac{2\lambda^i \, p_i \, p_0}{p_0 + p_4} - \frac{\mathbf{k}^2}{(p_0 + p_4)^2} \lambda^i \, p_i = \zeta^i \, S(p_i) \tag{6.33}$$

where we introduce a momentum-dependent infinitesimal parameter

$$\zeta^i = \lambda^i \, \frac{\kappa}{p_0 + p_4} \tag{6.34}$$

Thus, the Lorentz transformation of the zero component of the antipode is an ordinary Lorentz transformation, with parameter ζ^i.

For the spacial component we have a more complicated expression.

$$\begin{aligned}\delta_\lambda S(p_i) &= -\frac{\kappa \lambda_i \, p_0}{p_0 + p_4} + \frac{\kappa \, p_i}{(p_0 + p_4)^2} \lambda^j \, p_j \\ &= \zeta_i \, S(p_0) + \frac{\kappa}{(p_0 + p_4)^2} \left(p_i \lambda^j \, p_j - \lambda_i \mathbf{p}^2 \right)\end{aligned} \tag{6.35}$$

The first term here is again the standard Lorentz transformation with parameter ζ^i. The second term is an infinitesimal rotation of $S(p_i)$ with the parameter

$$\rho^j = \varepsilon^j{}_{kl} \, \lambda^k \, p^l$$

so that, finally

$$\delta_\lambda S(p_i) = \zeta_i \, S(p_0) + \varepsilon_i{}^{jk} \, \rho_j \, S(p_k) \tag{6.36}$$

Since under Lorentz boost transformation of momenta, the components of the antipode transform under a combination of boost and rotation, it is clear that the components of the antipode satisfy the same mass-shell condition as the components of the original momenta

$$S(p_0)^2 - S(p_i)^2 = m^2 = p_0^2 - p_i^2 \,.$$

In the case of the antipodal coordinates p_μ^*, the situation is similar. On-shell the condition (6.15) takes the form

$$p_0^* - \sqrt{m^2 + \kappa^2} < 0 \tag{6.37}$$

so that this time it does not impose any restrictions on negative energy states, but provides an upper bound on the positive energy ones $0 < p_0^* < \sqrt{m^2 + \kappa^2}$. Again one solves the apparent problem with Lorentz symmetry with the help of the antipode, which has the form

$$S(p_0^*) = -p_0^* + \frac{\mathbf{p}^{*2}}{p_0^* + p_4^*} = \frac{\kappa^2}{p_0^* + p_4^*} - p_4^*, \quad S(\mathbf{p}^*) = \frac{\kappa \mathbf{p}^*}{p_0^* + p_4^*}, \quad S(p_4^*) = p_4^*$$
(6.38)

On-shell $S(\omega_p^*) = S(-\sqrt{m^2 + \mathbf{p}^{*2}})$ is always positive.

6.1.2 Non-commutative Calculus

As in ordinary field theory, one expects the Casimir invariant (6.27) to have a 'coordinate space' counterpart in terms of a non-commutative wave operator. This will be written in terms of non-commutative differential operators complying with the non-trivial structure of the spacetime commutator (6.1) and of the symmetry generators. In this section, we introduce the differential calculus needed to define such operators.

It is well known [1,18] that it is impossible to construct a four-dimensional set of differentials which are also covariant under the action of κ-Poincaré generators. Rather, one has to resort to a five-dimensional set of differentials $\{d\hat{x}_0, d\hat{x}_1, d\hat{x}_2, d\hat{x}_3, d\hat{x}_4\}$ with the following commutation relations with the κ-Minkowski coordinates

$$[\hat{x}_0, d\hat{x}_0] = \frac{i}{\kappa} d\hat{x}_4 , \quad [\hat{x}_0, d\hat{x}_i] = 0 , \quad [\hat{x}_0, d\hat{x}_4] = \frac{i}{\kappa} d\hat{x}_0 ,$$

$$[\hat{x}_i, d\hat{x}_0] = \frac{i}{\kappa} d\hat{x}_i , \quad [\hat{x}_i, d\hat{x}_j] = \delta_{ij} \frac{i}{\kappa}(d\hat{x}_0 - d\hat{x}_4) , \quad [\hat{x}_i, d\hat{x}_4] = \frac{i}{\kappa} d\hat{x}_i . \quad (6.39)$$

and which commute among themselves

$$[d\hat{x}_A, d\hat{x}_B] = 0 , \quad A = 0, \ldots, 4$$
(6.40)

It should be noticed in passing that the fact that we have to do with five differentials corresponds naturally with the presence of five components of momenta $p_0, \ldots, p_4$.

It can be checked by taking the differential of both sides of (6.1) that these commutators are consistent with the non-commutative structure of spacetime and that all Jacobi identities involving differentials and non-commuting coordinates are satisfied. The Lorentz covariance of such relations can be easily checked using the relations [17]

$$N_i \triangleright \hat{x}_0 = i\hat{x}_i, \quad N_i \triangleright \hat{x}_j = i\delta_{ij}\hat{x}_0,$$
(6.41)

and extending the action to the differentials algebra in a natural way as[2]

$$N_i \triangleright d\hat{x}_\mu = d(N_i \triangleright \hat{x}_\mu) , \quad N_i \triangleright (\hat{x}_\mu d\hat{x}_\nu) = (N_i^{(1)} \triangleright \hat{x}_\mu)(d(N_i^{(2)} \triangleright \hat{x}_\nu)) , \quad (6.42)$$

---

[2] We also assume that the differential $d\hat{x}_4$ is κ-Poincaré invariant $\mathsf{p}_\kappa \triangleright d\hat{x}_4 = 0$, where p_κ is a generic element of the κ-Poincaré algebra.

where we have used Sweedler notation $\Delta(N) = N^{(1)} \otimes N^{(2)} = \sum_a N^{(1)a} \otimes N^{(2)a}$ for the coproduct in (6.28).

Instead of proving explicitly Lorentz covariance of the 5D non-commutative differentials algebra defined above, let us show that the 4D differential used in [5] and defined by the commutators

$$[\hat{x}_0, d\hat{x}_i] = -\frac{i}{\kappa} d\hat{x}_i \,, \quad [\hat{x}_0, d\hat{x}_0] = 0 \,, \quad [\hat{x}_i, d\hat{x}_\mu] = 0 \,, \tag{6.43}$$

fails to be κ-Lorentz covariant. To this end, consider

$$N_i \triangleright [\hat{x}_j, d\hat{x}_k] = \delta_{ij}[\hat{x}_0, d\hat{x}_k] = -\delta_{ij}\frac{i}{\kappa} d\hat{x}_i \neq N_i \triangleright 0 = 0$$

A differential on the algebra of functions over κ-Minkowski spacetime can be defined as

$$d = i d\hat{x}^A \hat{\partial}_A \,, \tag{6.44}$$

where the derivatives $\hat{\partial}_A$ are determined by requiring that the Leibniz rule for the differential is satisfied, as we now show. Working in the bicrossproduct basis, the explicit form of the $\hat{\partial}_A$ can be derived by first noting that, from the commutator $[\hat{x}_\mu, d\hat{x}_A] = (\hat{x}_\mu)^B_A d\hat{x}_B$, follows the identity

$$\hat{e}_k d\hat{x}^A \hat{e}_{S(k)} = d\hat{x}^B (\hat{e}_k)^A_B \,, \tag{6.45}$$

where the antipode in the bicrossproduct basis is defined in Eq. (6.20), $(\hat{e}_k)^A_B$ is the matrix representation (6.4) of the right-ordered plane wave $e^{ik_i \hat{x}^i} e^{ik_0 \hat{x}^0} \equiv \hat{e}_k$. Imposing then the Leibniz rule for the differential d on the product $\hat{e}_k \hat{e}_l$, we get

$$\begin{aligned} d(\hat{e}_k \hat{e}_l) &= (d\hat{e}_k)\hat{e}_l + \hat{e}_k(d\hat{e}_l) = (i d\hat{x}^A \hat{\partial}_A \hat{e}_k)\hat{e}_l + \hat{e}_k(i d\hat{x}^A \hat{\partial}_A \hat{e}_l) \\ &= i\big[(d\hat{x}^A \hat{\partial}_A \hat{e}_k)\hat{e}_l + \hat{e}_k(d\hat{x}^A \hat{e}_{S(p)} \hat{e}_k \hat{\partial}_A \hat{e}_l)\big] \\ &= i d\hat{x}^A \big[(\hat{\partial}_A \hat{e}_k)\hat{e}_l + (\hat{e}_k)^B_A \hat{e}_k(\hat{\partial}_b \hat{e}_l)\big] \,, \end{aligned} \tag{6.46}$$

where in the second term of the third equality, we have introduced $\hat{e}_{S(k)} \hat{e}_k = 1$ and in the last equality, we used the relation (6.45). Accordingly, looking at the first and the last terms of (6.46), we find that, in order for the differential $d = i d\hat{x}^A \hat{\partial}_A$ to satisfy the Leibniz rule, the derivatives must have coproducts

$$\Delta(\hat{\partial}_A) = \hat{\partial}_A \otimes 1 + (\hat{e}_k)^B_A \otimes \hat{\partial}_B \,. \tag{6.47}$$

It turns out that these coproducts reproduce the ones for the classical basis generators p_a in (6.25), with the coproduct of the operator $\hat{\partial}_4$ corresponding to the coproduct of $(\kappa - p_4)$. Therefore, we can identify non-commutative derivatives associated with the five-dimensional covariant calculus with translation generators of the classical basis. The action of the derivatives on right-ordered plane waves is then

$$\hat{\partial}_\mu \hat{e}_k = p_\mu(k)\hat{e}_k \,, \quad \hat{\partial}_4 \hat{e}_k = \big(\kappa - p_4(k)\big)\hat{e}_k \,, \tag{6.48}$$

where the explicit form of the classical basis momenta $p_A(k)$ in terms of the bicrossproduct momenta is given by (6.5). These derivatives form for us the desired differential calculus . One also defines conjugate operators $\hat{\partial}_\mu^\dagger$ whose action on plane waves is given by[3]

$$\hat{\partial}_\mu^\dagger \hat{e}_k \equiv (\hat{\partial}_\mu \hat{e}_k^\dagger)^\dagger = S\big(p(k)\big)_\mu \hat{e}_k \,, \quad \hat{\partial}_4^\dagger = \hat{\partial}_4 \,, \tag{6.49}$$

reflecting the fact that $p\big(S(k)\big)_A = S\big(p(k)\big)_A$.

From the action of the derivatives on $\hat{e}_k$, one can straightforwardly derive the action of the operators $\hat{\partial}_a$ on the generic function of non-commuting coordinates $\hat{f}(\hat{x})$. This can be done by resorting to the following Fourier expansion in terms of the non-commutative plane waves $\hat{e}_k$

$$\hat{f}(\hat{x}) = \int d\mu(k)\, \tilde{f}_r(k)\hat{e}_k(\hat{x}) \,, \tag{6.50}$$

where the integration measure $d\mu(k)$ is the Haar measure[4] on $AN(3)$

$$d\mu(k) = \frac{e^{3k_0/\kappa}}{(2\pi)^4} dk_0 d\mathbf{k} \,, \tag{6.51}$$

which can be also expressed in terms of the ordinary Lebesgue measure on the five-dimensional embedding space $d^5 p$ as

$$d^5\mu(p) = \kappa \frac{\delta(p_A p^A - \kappa^2)\theta(p_0 + p_4)}{(2\pi)^4} d^5 p \,, \tag{6.52}$$

Integrating over p_4, we obtain the measure expressed in terms of the four momenta p_μ that we will use below

$$d^4\mu(p) = \frac{\kappa}{p_4} \frac{1}{(2\pi)^4} \theta(p_0 + p_4)d^5 p \,, \tag{6.53}$$

with $p_4 \equiv \sqrt{p_0^2 - p^2 + \kappa^2}$ which on shell $p_0^2 - p^2 = m^2$ becomes just a constant $p_4 \equiv \sqrt{m^2 + \kappa^2}$

---

[3] Here we have used the fact that the hermitian conjugate of a plane wave involves the antipode map $S(k)$ on its momentum $\hat{e}_k^\dagger = \hat{e}_{S(k)}$.

[4] It is a left invariant measure $d\mu(lk) = d\mu(k)$, and it is worth noticing that in bicrossproduct coordinates it is just the diffeomorphism invariant measure on dS_4 corresponding to the cosmological metric $-dk_0^2 + e^{2k_0/\kappa} dk_i^2$.

6.1.3   Weyl Maps and ⋆-Product

The Weyl map is a useful tool first introduced in quantum mechanics to map classical observables (commuting functions on phase space) to quantum observables (functions of non-commuting operators). Due to ordering ambiguities on the non-commutative side, Weyl maps are obviously not unique. In our context, a Weyl map will map a function on commutative Minkowski spacetime to a (suitably ordered) function on the non-commutative κ-Minkowski spacetime.

Let us focus on plane waves. We define the right-ordered Weyl map $\mathscr{W}$ as

$$\mathscr{W}(e^{ipx}) = e^{ik_i\hat{x}^i}e^{ik_0\hat{x}^0} = \hat{e}_k, \quad \mathscr{W}^{-1}(\hat{e}_k) = e^{ipx}, \tag{6.54}$$

i.e., an ordinary plane wave is mapped to an $AN(3)$ group element written in the decomposition (6.3) in which the non-abelian generator $\hat{x}^0$ is always to the right. One can associate a commutative function $f(x)$ to $\hat{f}(\hat{x})$ via this Weyl map using the Fourier expansion (6.50)

$$f(x) = \mathscr{W}^{-1}(\hat{f}(\hat{x})) = \int d\mu(p)\tilde{f}(p)\mathscr{W}^{-1}(\hat{e}_k(\hat{x})) = \int d\mu(p)\tilde{f}(p)e^{ipx}. \tag{6.55}$$

The definition of Weyl map is not unambiguous, and the ambiguity can be resolved by assuming that the map leads to non-commutative plane waves on which the derivatives $\hat{\partial}_\mu$ of the non-commutative differential calculus have 'classical' action, i.e.

$$\hat{\partial}_\mu \triangleright \mathscr{W}(e^{ipx}) = \mathscr{W}(-i\partial_\mu e^{ipx}) = p_\mu \mathscr{W}(e^{ipx}). \tag{6.56}$$

The Weyl map $\mathscr{W}$ is related to the classical basis coordinates p_A, and as it can be easily checked confronting the actions (6.48) and (6.56), it has the following action on plane waves:

$$\mathscr{W}(e^{ipx}) = \hat{e}_{k(p)}, \quad \mathscr{W}^{-1}(\hat{e}_{k(p)}) = e^{ipx}, \tag{6.57}$$

i.e., $\mathscr{W}$ maps a commutative plane wave labelled by p to a right-ordered non-commutative plane wave whose four-momentum is $p_\mu(p)$, where $p(p)$ is the inverse transformation of (6.5). Therefore, following (6.50), a non-commutative function $\hat{f}(\hat{x})$ can be expressed as

$$\hat{f}(\hat{x}) = \int d\mu(p)\tilde{f}(p)\mathscr{W}(e^{ipx}), \tag{6.58}$$

where $\tilde{f}(p) = \tilde{f}(k(p))$, and the commutative function $f(x)$ associated to $\hat{f}(\hat{x})$ through the inverse classical basis Weyl map is given by

$$f(x) = \mathscr{W}^{-1}(\hat{f}(\hat{x})) = \int d\mu(p)\tilde{f}(p)e^{ipx}. \tag{6.59}$$

Using such map, we can finally introduce a suitable notion of integration on κ-Minkowski space as follows:

$$\widehat{\int} \hat{f} = \int d^4x \; \mathscr{W}^{-1}(\hat{f}(\hat{x})) = \int d^4x \; f(x) \,. \tag{6.60}$$

On the space of commutative functions obtained via the action of $\mathscr{W}^{-1}$, the non-commutativity of κ-Minkowski space is reflected in a non-trivial $\star$-product which replaces the ordinary commutative pointwise product. The star product associated to the Weyl map $\mathscr{W}$ is defined by the relation

$$\hat{f}(\hat{x})\hat{g}(\hat{x}) = \mathscr{W}(f(x))\mathscr{W}(g(x)) = \mathscr{W}(f(x)\star g(x)) \,. \tag{6.61}$$

The explicit expressions for the star product can be found in [20].

6.1.4 Summary of the Technical Tools

In this section, we will summarize all the tools needed to construct the complex scalar field theory. The field will be constructed both in spacetime and momentum space; we have therefore two concepts $\phi(x)$ and $\phi(p)$ that are going to be related by Fourier transform. The non-commutative structure of κ-Minkowski space is then taken into account by the assumption that the multiplication of the field is given by star product that we have constructed above.

Since we assume that the spacetime and momentum space fields of interest are related by Fourier transform it is sufficient to define the star product and differential calculus for plane waves only. Following [1], we define the Weyl map $\mathscr{W}$ that maps group elements (plane waves on non-commutative κ-Minkowski spacetime) to ordinary plane waves on commutative spacetime manifold with coordinates x

$$\mathscr{W}(\hat{e}_k(\hat{x})) = e_p(x) \tag{6.62}$$

defined by the action of the derivatives

$$\mathscr{W}(\hat{\partial}_\mu \hat{e}_k)(\hat{x}) = \partial_\mu e_p(x) \,, \quad \mathscr{W}(\hat{\partial}_\mu \hat{e}_k^*)(\hat{x}) = \partial_\mu e_p^*(x) \tag{6.63}$$

with ∂_μ being the standard partial derivative. As a result, we have

$$e_p(x) = e^{ip_\mu x^\mu} = e^{-i(\omega_\mathbf{p} t - \mathbf{p}\mathbf{x})} \,, \quad e_p^*(x) = e^{ip_\mu^* x^\mu} = e^{-i(\omega_\mathbf{p}^* t - \mathbf{p}^*\mathbf{x})} \tag{6.64}$$

with the on-shell relations

$$\omega_\mathbf{p} = \sqrt{m^2 + p^2}\,, \quad \omega_\mathbf{p}^* = -\sqrt{m^2 + p^{*2}}\,, \quad p_4 = \sqrt{m^2 + \kappa^2}\,, \quad p_4^* = -\sqrt{m^2 + \kappa^2} \tag{6.65}$$

Using the Weyl map we construct the star product of two commuting plane waves from the product of two group elements

$$\mathscr{W}(\hat{e}_k\,\hat{e}_l) \equiv e_{p(k)} \star e_{q(l)} = e_{p \oplus q} \tag{6.66}$$

In the case of two positive energy plane waves, we have

$$\hat{e}_k\,\hat{e}_l = \hat{e}_{k \oplus l} \tag{6.67}$$

with

$$(k \oplus l)_0 = k_0 + l_0, \quad (k \oplus l)_i = k_i + e^{-k_0/\kappa}\,l_i \tag{6.68}$$

Acting on the reference vector $(0, \ldots, 0, \kappa)$ with the group element (6.67) we derive the momentum composition rule

$$
\begin{aligned}
(p \oplus q)_0 &= \frac{1}{\kappa}\,p_0(q_0 + q_4) + \frac{\mathbf{p}\mathbf{q}}{p_0 + p_4} + \frac{\kappa}{p_0 + p_4}\,q_0 \\
(p \oplus q)_i &= \frac{1}{\kappa}\,p_i(q_0 + q_4) + q_i \\
(p \oplus q)_4 &= \frac{1}{\kappa}\,p_4(q_0 + q_4) - \frac{\mathbf{p}\mathbf{q}}{p_0 + p_4} - \frac{\kappa}{p_0 + p_4}\,q_0
\end{aligned}
\tag{6.69}
$$

which is in one-to-one correspondence with the coproducts (6.24).

We can use the same construction in the case of the negative energy plane waves, corresponding to the antipodal coordinates p^*. We first compute the product

$$\vartheta\,e_{(p_0, \mathbf{p})} = e_{(p_0, -\mathbf{p})}\,\vartheta \tag{6.70}$$

and then using

$$\mathscr{W}(\hat{e}_k^*\,\hat{e}_l) \equiv e_{p(k)}^* \star e_{q(l)} = e_{p^* \oplus q} \tag{6.71}$$

we obtain

$$
\begin{aligned}
(p^* \oplus q)_0 &= \frac{1}{\kappa}\,p_0^*(q_0 + q_4) + \frac{\mathbf{p}^*\mathbf{q}}{p_0^* + p_4^*} + \frac{\kappa}{p_0^* + p_4^*}\,q_0 \\
(p^* \oplus q)_i &= \frac{1}{\kappa}\,p_i^*(q_0 + q_4) + q_i \\
(p^* \oplus q)_4 &= \frac{1}{\kappa}\,p_4^*(q_0 + q_4) - \frac{\mathbf{p}^*\mathbf{q}}{p_0^* + p_4^*} - \frac{\kappa}{p_0^* + p_4^*}\,q_0
\end{aligned}
\tag{6.72}
$$

(In the computation one takes (6.69), changes the overall sign, then changes the sign of p replacing it by p^*, and finally changes the sign of $\mathbf{q}$ according to (6.70).)

Analogously

$$(p \oplus q^*)_0 = \frac{1}{\kappa} p_0 (q_0^* + q_4^*) + \frac{\mathbf{p}\mathbf{q}^*}{p_0 + p_4} + \frac{\kappa}{p_0 + p_4} q_0^*$$

$$(p \oplus q^*)_i = \frac{1}{\kappa} p_i (q_0^* + q_4^*) + q_i^*$$

$$(p \oplus q^*)_4 = \frac{1}{\kappa} p_4 (q_0^* + q_4^*) - \frac{\mathbf{p}\mathbf{q}^*}{p_0 + p_4} - \frac{\kappa}{p_0 + p_4} q_0^* \qquad (6.73)$$

Finally, we consider the composition of two negative energy plane waves. (In this case, after moving through the q plane wave, we get $\vartheta^2 = 1$)

$$(p^* \oplus q^*)_0 = \frac{1}{\kappa} p_0^* (q_0^* + q_4^*) + \frac{\mathbf{p}^*\mathbf{q}^*}{p_0^* + p_4^*} + \frac{\kappa}{p_0^* + p_4^*} q_0^*$$

$$(p^* \oplus q^*)_i = \frac{1}{\kappa} p_i^* (q_0^* + q_4^*) + q_i^*$$

$$(p^* \oplus q^*)_4 = \frac{1}{\kappa} p_4^* (q_0^* + q_4^*) - \frac{\mathbf{p}^*\mathbf{q}^*}{p_0^* + p_4^*} - \frac{\kappa}{p_0^* + p_4^*} q_0^* \qquad (6.74)$$

Notice that, remarkably, all the composition laws (6.69)–(6.74) have exactly the same form so there is no need to distinguish between them.

Let us now discuss the conjugation of our non-commutative plane waves. In the undeformed case, we define the conjugation of a plane wave as a solution of the equations $(e^{ipx})^\dagger e^{ipx} = 1 = e^{ipx} (e^{ipx})^\dagger$ from which it immediately follows that $(e^{ipx})^\dagger = e^{-ipx}$. In the deformed case, we proceed analogously, demanding that

$$e_p^\dagger \star e_p = 1 = e_p \star e_p^\dagger . \qquad (6.75)$$

The solution of this equation inolves the antipode (6.25)

$$\left(e^{-ipx} \right)^\dagger = e^{-iS(p)x} . \qquad (6.76)$$

Notice that if e^{-ipx} is a positive energy plane wave then $\left(e^{-ipx} \right)^\dagger = e^{-iS(p)x}$ is the negative energy one.

Before ending this summary, we discuss the integration that we will need to write down the field action. The integral over non-commutative κ-Minkowski spacetime (6.1), $\hat{\mathbb{R}}^4$ is related to the integral on commutative Minkowski spacetime $\mathbb{R}^4$ as follows:

$$\int_{\hat{\mathbb{R}}^4} \hat{\phi}(\hat{x}) = \int_{\mathbb{R}^4} d^4 x \, \phi(x) , \qquad \int_{\hat{\mathbb{R}}^4} \hat{\phi}(\hat{x}) \, \hat{\psi}(\hat{x}) = \int_{\mathbb{R}^4} d^4 x \, \phi(x) \star \psi(x) \qquad (6.77)$$

This integral is not cyclic, because

$$\int_{\hat{\mathbb{R}}^4} \hat{e}_k \hat{e}_l = e^{3k_0} \int_{\hat{\mathbb{R}}^4} \hat{e}_l \hat{e}_k \qquad (6.78)$$

but it satisfies the exchange property

$$\int_{\hat{\mathbb{R}}^4} \hat{e}_k^\dagger \hat{e}_l = \int_{\hat{\mathbb{R}}^4} \hat{e}_l^\dagger \hat{e}_k \tag{6.79}$$

or

$$\int_{\mathbb{R}^4} d^4x\, e_p^\dagger \star e_q = \int_{\mathbb{R}^4} d^4x\, e_q^\dagger \star e_p \tag{6.80}$$

We will make use of these properties in what follows.

6.2   Action and Field Equations

In this section, we will consider a deformed free complex scalar field $\phi(x)$, which we will assume to satisfy the standard Klein-Gordon equation as its field equation

$$\left(\partial_\mu \partial^\mu - m^2\right)\phi(x) = 0 \tag{6.81}$$

and its adjoint. Since the field equations are linear, the action should be linear in the field $\phi(x)$ and its adjoint $\phi^\dagger(x)$ as well. In this section, we will construct the field action and derive the field equations. With the action at hands we will also be able then to construct conserved quantities as well as the symplectic structure. The latter will be the starting point for the proposal of a deformed algebra of creation and annihilation operators.

Taking into account the exchange property of the integral, we see that the most general form of deformed free complex field action has the form

$$S = \frac{1}{2} \int_{\mathbb{R}^4} d^4x \left[(\partial_\mu \phi)^\dagger \star \partial^\mu \phi + (\partial_\mu \phi) \star (\partial^\mu \phi)^\dagger + m^2 (\phi^\dagger \star \phi + \phi \star \phi^\dagger) \right] \tag{6.82}$$

To derive the field equations, we must learn how to integrate by parts. The starting point is provided by the coproduct rules for the κ-Poincaré algebra in the classical basis (p_0, p_i, p_4) (6.24).

The coproducts tell us how the momentum operators act on star products of two functions. Since momenta are spacetime derivatives $p_0 = i\partial_0$, $p_i = i\partial_i$ these equations tell us how derivatives act on the star products of functions on Minkowski space, defining in this way the deformed Leibniz rule. In the calculation below, we use the short-hand notation $p_+ \to \Delta_+ = i\partial_0 + p_4 = i\partial_0 + (\kappa + i\partial_4)$, where the nonlocal operator p_4 is expressed in terms of the corresponding derivatives as $p_4 = \sqrt{\kappa^2 - \partial_0^2 + \partial_i^2}$. Equation (6.24) then imply

$$\partial_0(\phi \star \psi) = \frac{1}{\kappa}(\partial_0 \phi) \star (\Delta_+ \psi) + \kappa(\Delta_+^{-1}\phi) \star (\partial_0 \psi) + i(\Delta_+^{-1}\partial_i \phi) \star (\partial_i \psi) \tag{6.83}$$

$$\partial_i(\phi \star \psi) = \frac{1}{\kappa}(\partial_i \phi) \star (\Delta_+ \psi) + \phi \star (\partial_i \psi) \tag{6.84}$$

$$\Delta_+(\phi \star \psi) = \frac{1}{\kappa}(\Delta_+ \phi) \star (\Delta_+ \psi). \tag{6.85}$$

Defining the adjoint derivative

$$(\partial_A \phi)^\dagger \equiv \partial_A^\dagger \phi^\dagger, \quad A = (\mu, 4, +) \tag{6.86}$$

and using Eq. (6.24), we have

$$\partial_i^\dagger = \kappa \Delta_+^{-1} \partial_i, \quad \partial_0^\dagger = \partial_0 - i \Delta_+^{-1} \partial^2, \quad \partial_4^\dagger = -\partial_4, \quad \Delta_+^\dagger = \kappa^2 \Delta_+^{-1}. \tag{6.87}$$

We now use Eqs. (6.83), (6.84), (6.85), (6.87) to obtain the expressions needed for the integration by parts of expressions of the form $(\partial_\mu \phi)^\dagger \star \partial^\mu \psi$ and $(\partial_\mu \psi) \star (\partial^\mu \phi)^\dagger$. With some algebra, we find

$$(\partial_i \phi)^\dagger \star (\partial_i \psi) = \partial_i \left[(\partial_i \phi)^\dagger \star \psi \right] - \frac{\Delta_+}{\kappa} \left[(\partial^2 \phi)^\dagger \star \psi \right]. \tag{6.88}$$

Similarly

$$(\partial_0 \phi)^\dagger \star (\partial_0 \psi) = \frac{\partial_0}{\kappa} \left[(\Delta_+ (\partial_0 \phi)^\dagger) \star \psi \right] - i \partial_i [(\Delta_+^{-1} \partial_i \partial_0 \phi)^\dagger \star \psi] - \frac{\Delta_+}{\kappa} \left[(\partial_0^2 \phi)^\dagger \star \psi \right]. \tag{6.89}$$

The conjugates of Eqs. (6.88), (6.89) take the form

$$(\partial_i \psi)^\dagger \star (\partial_i \phi) = \partial_i^\dagger \left[\psi^\dagger \star (\partial_i \phi) \right] - \frac{\kappa}{\Delta_+} \left[\psi^\dagger \star (\partial^2 \phi) \right] \tag{6.90}$$

$$(\partial_0 \psi)^\dagger \star (\partial_0 \phi) = \partial_0^\dagger \left[\psi^\dagger \star (\kappa \Delta_+^{-1} \partial_0 \phi) \right] + i \partial_i^\dagger [\psi^\dagger \star (\Delta_+^{-1} \partial_i \partial_0 \phi)] - \frac{\kappa}{\Delta_+} \left[\psi \star (\partial_0^2 \phi) \right]. \tag{6.91}$$

Finally, we will also need the following identity:

$$m^2 \phi^\dagger \star \psi = -\left(\frac{\Delta_+}{\kappa} - 1 \right) (m^2 \phi^\dagger \star \psi) + \frac{\Delta_+}{\kappa} (m^2 \phi^\dagger \star \psi)$$

$$= -\frac{i \partial_0}{\kappa} (m^2 \phi^\dagger \star \psi) - \frac{i \partial_4}{\kappa} (m^2 \phi^\dagger \star \psi) + \frac{\Delta_+}{\kappa} (m^2 \phi^\dagger \star \psi) \tag{6.92}$$

and its conjugate

$$m^2 \psi^\dagger \star \phi = +\frac{i \partial_0^\dagger}{\kappa} (m^2 \psi^\dagger \star \phi) + \frac{i \partial_4^\dagger}{\kappa} (m^2 \psi^\dagger \star \phi) + \frac{\kappa}{\Delta_+} (m^2 \psi^\dagger \star \phi). \tag{6.93}$$

For the opposite ordering, we have instead

$$(\partial_i \psi) \star (\partial_i \phi)^\dagger = \kappa \partial_i (\psi \star [\Delta_+^{-1} (\partial_i \phi)^\dagger]) - \psi \star (\partial^2 \phi)^\dagger. \tag{6.94}$$

$$(\partial_0 \psi) \star (\partial_0 \phi)^\dagger = \partial_0 (\psi \star [\kappa \Delta_+^{-1} (\partial_0 \phi)^\dagger]) - i \partial_i (\psi \star [\Delta_+^{-1} \partial_i (\partial_0 \phi)^\dagger]) - [\psi \star (\partial_0^2 \phi)^\dagger]$$

$$+ \left(\frac{i}{\kappa} \partial_0 + i \frac{\partial_4}{\kappa} \right) [\psi \star (\partial_0^2 \phi)^\dagger]. \tag{6.95}$$

We can now compute the variation of the action in Eq. (6.82). Writing $S = S_1 + S_2$, where

$$S_1 = \frac{1}{2} \int_{\mathbb{R}^4} d^4 x \ (\partial^\mu \phi)^\dagger \star (\partial_\mu \phi) + m^2 \phi^\dagger \star \phi \tag{6.96}$$

$$S_2 = \frac{1}{2} \int_{\mathbb{R}^4} d^4 x \ (\partial_\mu \phi) \star (\partial^\mu \phi)^\dagger + m^2 \phi \star \phi^\dagger. \tag{6.97}$$

we have

$$\delta S_1 = \frac{1}{2} \int_{\mathbb{R}^4} d^4 x \ (\partial_\mu \delta \phi)^\dagger \star \partial^\mu \phi + (\partial_\mu \phi)^\dagger \star \partial^\mu \delta \phi + m^2 \delta \phi^\dagger \star \phi + m^2 \phi^\dagger \star \delta \phi \tag{6.98}$$

which can be rewritten as

$$\delta S_1 = \frac{1}{2} \int_{\mathbb{R}^4} d^4 x \left\{ -\frac{\Delta_+}{\kappa} \left[(\partial_\mu^\dagger (\partial^\mu)^\dagger - m^2) \phi^\dagger \star \delta \phi \right] + \partial_A \left(\Pi_1^A \star \delta \phi \right) \right.$$

$$\left. - \frac{\kappa}{\Delta_+} \left[\delta \phi^\dagger \star (\partial_\mu \partial^\mu - m^2) \phi \right] + \partial_A^\dagger \left(\delta \phi^\dagger \star \left(\Pi_1^A \right)^\dagger \right) \right\} \tag{6.99}$$

where

$$\Pi_1^0 = (\Pi_0)_1 = \frac{1}{\kappa} (\Delta_+ \partial_0^\dagger + im^2) \phi^\dagger \tag{6.100}$$

$$\Pi_1^i = -(\Pi_i)_1 = (-\partial_i (1 + i \Delta_+^{-1} \partial_0)) \phi^\dagger \tag{6.101}$$

$$\Pi_1^4 = (\Pi_4)_1 = -i \frac{m^2 \phi^\dagger}{\kappa} \tag{6.102}$$

Analogously, we find

$$\delta S_2 = \frac{1}{2} \int_{\mathbb{R}^4} d^4 x \ \partial^\mu \phi \star (\partial_\mu \delta \phi)^\dagger + \partial^\mu \delta \phi \star (\partial_\mu \phi)^\dagger + m^2 \phi \star \delta \phi^\dagger + m^2 \delta \phi \star \phi^\dagger \tag{6.103}$$

which can be rewritten as

$$\delta S_2 = \frac{1}{2} \int_{\mathbb{R}^4} d^4x \left\{ -\left[\delta\phi \star (\partial_\mu^\dagger (\partial^\mu)^\dagger - m^2)\phi^\dagger\right] + \partial_A \left(\delta\phi \star \Pi_2^A\right) \right.$$

$$\left. - \left[(\partial_\mu \partial^\mu - m^2)\phi \star \delta\phi^\dagger\right] + \partial_A^\dagger \left(\left(\Pi_2^A\right)^\dagger \star \delta\phi^\dagger\right) \right\} \quad (6.104)$$

where

$$\Pi_2^0 = (\Pi_0)_2 = \left(\frac{\kappa}{\Delta_+} \partial_0^\dagger + \frac{i}{\kappa}(\partial_0^\dagger)^2\right)\phi^\dagger \quad (6.105)$$

$$\Pi_2^i = -(\Pi_i)_2 = -\frac{\kappa}{\Delta_+}(\partial_i^\dagger + i\partial_i\partial_0^\dagger)\phi^\dagger \quad (6.106)$$

$$\Pi_2^4 = (\Pi_4)_2 = +i\frac{(\partial_0^\dagger)^2}{\kappa}\phi^\dagger. \quad (6.107)$$

Therefore, the field equations have the expected form

$$(\partial_\mu \partial^\mu - m^2)\phi = 0, \quad (\partial_\mu^\dagger (\partial^\mu)^\dagger - m^2)\phi^\dagger = 0 \quad (6.108)$$

As we will see below, these equations lead to two nontrivially related mass-shell conditions, describing the same orbit of the Lorentz group.

6.3 Complex Scalar Field and Momentum Space Action

In the previous section, we derive the complex scalar field equations without specifying what the scalar field actually is, i.e., without fixing its Fourier decomposition in the plane wave basis. Here we will complete the construction, postulating the specific form of the field, which makes the invariance under continuous and discrete symmetries easy to trace.

Recalling that $(e^{ipx})^\dagger = e^{iS(p)x}$, we postulate the following form of the field:

$$\phi(x) = \int \frac{d^3p}{\sqrt{2\omega_p}} \left[1 + \frac{|p_+|^3}{\kappa^3}\right]^{-\frac{1}{2}} a_{\mathbf{p}}\, e^{-i(\omega_p t - \mathbf{px})}$$

$$+ \int \frac{d^3p^*}{\sqrt{2|\omega_p^*|}} \left[1 + \frac{|p_+^*|^3}{\kappa^3}\right]^{-\frac{1}{2}} b_{\mathbf{p}^*}^\dagger\, e^{i(S(\omega_p^*)t - S(\mathbf{p}^*)x)}$$

$$\equiv \phi_{(+)}(x) + \phi_{(-)}(x) \quad (6.109)$$

where we include an additional factor $(1 + |p_+|^3/\kappa^3)^{-1/2}$ (which can be regarded as a modification of the integration measure) because it makes the form of the momentum space action, which we will construct in a moment particularly simple.

The adjoint field is

$$\phi^\dagger(x) = \int \frac{d^3p}{\sqrt{2\omega_p}} \left[1 + \frac{|p_+|^3}{\kappa^3}\right]^{-\frac{1}{2}} a_{\mathbf{p}}^\dagger\, e^{-i(S(\omega_p)t - S(\mathbf{p})x)}$$

$$+ \int \frac{d^3p^*}{\sqrt{2|\omega_p^*|}} \left[1 + \frac{|p_+^*|^3}{\kappa^3}\right]^{-\frac{1}{2}} b_{\mathbf{p}^*}\, e^{i(\omega_p^* t - \mathbf{p}^* \mathbf{x})}$$

$$\equiv \phi_{(+)}^\dagger(x) + \phi_{(-)}^\dagger(x) \tag{6.110}$$

Since $\omega_p > 0$ and $S(\omega_p^*) > 0$ the field (6.109) is a combination of positive energy particle states and negative energy antiparticle ones, while in (6.109) we have the opposite arrangement, as it should be.

Let us plug these expressions to the action (6.82). We first split it in two parts, with a different ordering of the factors with respect to the star product, i.e., $S = \frac{1}{2}(S_1 + S_2)$, where

$$S_1 = \int_{\mathbb{R}^4} d^4x\, (\partial_\mu \phi)^\dagger \star \partial^\mu \phi + m^2 \phi^\dagger \star \phi \tag{6.111}$$

$$S_2 = \int_{\mathbb{R}^4} d^4x\, (\partial_\mu \phi) \star (\partial^\mu \phi)^\dagger + m^2 \phi \star \phi^\dagger \tag{6.112}$$

Let us start with S_1. For simplicity, in the course of calculations, we will denote $\left[1 + \frac{|p_+|^3}{\kappa^3}\right]^{-\frac{1}{2}} := \xi(p)$. Substituting (6.109) and (6.110) to (6.111), we get

$$S_1 = \int d^4x \int \frac{d^3p}{\sqrt{2\omega_p}} \frac{d^3q}{\sqrt{2\omega_q}} \Big\{ (p_\mu q^\mu + m^2)\xi(p)\xi(q) a_{\mathbf{p}}^\dagger a_{\mathbf{q}} e^{-i(S(\omega_p)\oplus\omega_q)t} e^{i(S(\mathbf{p})\oplus\mathbf{q})x}$$

$$\tag{6.113}$$

$$- (p_\mu S(q^*)^\mu + m^2)\xi(p)\xi(q) a_{\mathbf{p}}^\dagger b_{\mathbf{q}^*}^\dagger e^{-i(S(\omega_p)\oplus S(\omega_q))t} e^{i(S(\mathbf{p})\oplus S(\mathbf{q}))x}$$

$$\tag{6.114}$$

$$- (S(p^*)_\mu q^\mu + m^2)\xi(p)\xi(q) b_{\mathbf{p}^*} a_{\mathbf{q}} e^{-i(\omega_p \oplus \omega_q)t} e^{i(\mathbf{p}\oplus\mathbf{q})x} \tag{6.115}$$

$$+ (S(p^*)_\mu S(q^*)^\mu + m^2)\xi(p)\xi(q) b_{\mathbf{p}^*} b_{\mathbf{q}^*}^\dagger e^{-i(\omega_p \oplus S(\omega_q))t} e^{-i(\mathbf{p}\oplus S(\mathbf{q}))x} \Big\}$$

$$\tag{6.116}$$

The terms (6.114) and (6.115) vanish identically because $S(\omega_p) \oplus S(\omega_q) < 0$ and $\omega_p \oplus \omega_q > 0$. Let us call S_I and S_{II} the terms (6.113) and (6.116), respectively.

In the first term, we have to do with the expression $S(\mathbf{p}) \oplus \mathbf{q} = -\mathbf{p}\frac{q\pm}{p+} + \mathbf{q}$. After changing variables $\mathbf{P} = \mathbf{p}\frac{q\pm}{p+}$, the first term becomes

$$S_I = \int \frac{d^3 P}{\sqrt{2\omega_p(P, q)}} \frac{d^3 q}{\sqrt{2\omega_q}} (p_\mu(P)q^\mu + m^2)$$
$$a^\dagger_{\mathbf{p}(P)} a_\mathbf{q} \xi(p(P)) \xi(q) \delta(S(\omega_p(P)) \oplus \omega_q) \delta(\mathbf{q} - \mathbf{P}) |\det J| \tag{6.117}$$

where all the quantities that contained p now contain $p(P)$. Computing q the integral and changing back coordinates using the inverse relation of $\mathbf{P} = \mathbf{p}\frac{q\pm}{p+}$, we get a $\det J^{-1}$ we see that the determinant cancels, leaving us with

$$S_I = \int \frac{d^3 p}{\sqrt{2\omega_p}} \frac{1}{\sqrt{2\omega_P(p)}} (p_\mu P^\mu(p) + m^2) a^\dagger_\mathbf{p} a_{\mathbf{P}(p)} \xi(p) \xi(P(p)) \delta(S(\omega_p) \oplus \omega_{P(p)}) \tag{6.118}$$

Since p and q are on-shell

$$\omega_p^2 - \mathbf{p}^2 = m^2 = \omega_q^2 - \mathbf{q}^2 \tag{6.119}$$

after making use of the Dirac delta $\delta(\mathbf{q} - \mathbf{P})$ we find

$$\omega_p^2 - \mathbf{p}^2 = \omega_q^2 - \mathbf{p}^2 \frac{q_+^2}{p_+^2} = \omega_q^2 - \mathbf{p}^2 \frac{(\omega_q + \sqrt{\kappa^2 + m^2})^2}{(\omega_p + \sqrt{\kappa^2 + m^2})^2} \tag{6.120}$$

or equivalently

$$(\omega_p - \omega_q)(\omega_p + \omega_q) = \mathbf{p}^2 \frac{(\omega_p - \omega_q)(\omega_p + \omega_q + 2\sqrt{\kappa^2 + m^2})}{(\omega_p + \sqrt{\kappa^2 + m^2})^2}. \tag{6.121}$$

This equality is satisfied if $\omega_p = \omega_q$. If, however, we assume that $\omega_p \neq \omega_q$, then we must have

$$(\omega_p + \omega_q)(\omega_p + \sqrt{\kappa^2 + m^2})^2 = (\omega_p^2 - m^2)(\omega_p + \omega_q + 2\sqrt{\kappa^2 + m^2}). \tag{6.122}$$

Performing the products, this relation reduces to

$$\omega_p \kappa^2 + \omega_p m^2 + \omega_q \kappa^2 + \omega_q m^2 + 2\omega_q \omega_p \sqrt{\kappa^2 + m^2}$$
$$= -m^2 \omega_p - m^2 \omega_q - 2m^2 \sqrt{\kappa^2 + m^2} \tag{6.123}$$

which is a contradiction because the LHS is positive and the RHS negative. Therefore, $\omega_p = \omega_q$, and as a consequence $p_+ = q_+$ and hence $\mathbf{P} = \mathbf{p}$. Therefore, the first term of the action becomes simply

$$S_I = \int \frac{d^3p}{2\omega_p}(p_\mu p^\mu + m^2)a_\mathbf{p}^\dagger a_\mathbf{p}\xi^2(p)\delta(0) \tag{6.124}$$

In this expression, the formally infinite overall multiplicative factor $\delta(0)$ appears. Its emergence reflects the fact that in passing from the spacetime action to the momentum space one we have to do the integration over time of a time-independent quantity and the time axis has infinite length. This infinite constant does not change the variational properties of the action, and therefore, can be neglected. Furthermore, we are writing the creation/annihilation 'operators' in normal ordering, even though we still have to do with the classical theory, in which $a_\mathbf{p}$, $a_\mathbf{p}^\dagger$ and $b_\mathbf{p}$, $b_\mathbf{p}^\dagger$ are just commuting functions.

Similarly, in S_{II} we have

$$\mathbf{p} \oplus S(\mathbf{q}) = \frac{\kappa}{q_+}(\mathbf{p} - \mathbf{q}) \tag{6.125}$$

so that the second term of the action becomes

$$S_{II} = \int \frac{d^3p}{\sqrt{2\omega_p}}\frac{d^3q}{\sqrt{2\omega_q}}(S(p^*)_\mu S(q^*)^\mu + m^2)$$

$$\xi(p)\xi(q)b_{\mathbf{p}^*}b_{\mathbf{q}^*}^\dagger\delta(\omega_p \oplus S(\omega_q))\delta\left(\frac{\kappa}{q_+}(\mathbf{p} - \mathbf{q})\right) \tag{6.126}$$

In this case, computing first the integral with respect to p, one immediately gets

$$S_{II} = \int \frac{d^3p}{2\omega_p}\xi^2(q)(S(q)_\mu S(q)^\mu + m^2)b_{\mathbf{q}^*}b_{\mathbf{q}^*}^\dagger\frac{|q_+|^3}{\kappa^3}\delta(0) \tag{6.127}$$

The final form of the action S_1 in momentum space is, therefore,

$$S_1 = \int \frac{d^3p}{2\omega_p}\xi^2(p)\left[(p_\mu p^\mu + m^2)a_\mathbf{p}^\dagger a_\mathbf{p} + (S(p)_\mu S(p)^\mu + m^2)b_{\mathbf{p}^*}b_{\mathbf{p}^*}^\dagger\frac{|p_+|^3}{\kappa^3}\right]. \tag{6.128}$$

Concerning the action S_2, the computations proceed in exactly the same way as before, with the only exception that the Dirac deltas are switched. More explicitly, the exponential multiplying the term $a_\mathbf{q}^\dagger a_\mathbf{p}$ in S_2 is the same that multiplied the term $b_{\mathbf{p}^*}^\dagger b_{\mathbf{q}^*}$ in S_1, and vice versa. Therefore, we have

$$S_2 = \int \frac{d^3p}{2\omega_p}\xi^2(p)\left[\frac{|p_+|^3}{\kappa^3}(p_\mu p^\mu + m^2)a_\mathbf{p}^\dagger a_\mathbf{p} + (S(p)_\mu S(p)^\mu + m^2)b_{\mathbf{p}^*}b_{\mathbf{p}^*}^\dagger\right]. \tag{6.129}$$

Finally, the total action is given by

$$S = \frac{1}{2} \int \frac{d^3 p}{2\omega_p} \left[(p_\mu p^\mu + m^2) a_{\mathbf{p}}^\dagger a_{\mathbf{p}} + (S(p)_\mu S(p)^\mu + m^2) b_{\mathbf{p}^*} b_{\mathbf{p}^*}^\dagger \right]. \tag{6.130}$$

where we used the fact that

$$\xi^2(p) \left(1 + \frac{|p_+|^3}{\kappa^3} \right) = \left(1 + \frac{|p_+|^3}{\kappa^3} \right)^{-1} \left(1 + \frac{|p_+|^3}{\kappa^3} \right) = 1 \tag{6.131}$$

It follows from the variation of the action (6.130) with respect to the mode functions $a_{\mathbf{q}}, a_{\mathbf{p}}^\dagger, b_{\mathbf{p}^*}, b_{\mathbf{q}^*}^\dagger$ that the field equations for the 'particle' modes $a_{\mathbf{q}}^\dagger \, a_{\mathbf{p}}$ have the form of the mass-shell condition

$$p_\mu p^\mu = -m^2 \tag{6.132}$$

while for the 'antiparticle' modes $b_{\mathbf{p}^*}, b_{\mathbf{q}^*}^\dagger$ we obtain

$$S(p)_\mu S(p)^\mu = -m^2 \tag{6.133}$$

These two conditions define the same three-dimensional submanifold of momentum space, but are different and we can expect that they point to some subtle difference between (deformed) particles and antiparticles. In the next chapter, we will see that this is indeed the case.

Above we computed the momentum space action using on-shell fields. The resulting action is, strictly speaking, identically zero, because on-shell the conditions (6.132) and (6.133) are identically satisfied. However it still make sense to investigate the invariance properties of the action (6.130), which we will do in the following section. An alternative construction using the off-shell fields is presented in the paper [3].

6.4 Poincaré Symmetry of the Action

In this section, we conclude the investigations of classical κ-deformed complex scalar field by showing that the action is Poincaré-invariant, and therefore, we are indeed dealing with a properly defined relativistic field theory.

We take as a starting point the momentum space action (6.130), which, as we have shown above is equivalent to the original spacetime action (6.82) for the fields (6.109), (6.110). We choose to use the momentum space action and not the spacetime one here because in the case of the former the proof of Poincaré invariance is much simpler. For the complementary proof in spacetime see [3].

Poincaré group consists of translations and Lorentz transformations and let us consider the simpler case of translations first. In order to do so, we must first find out how spacetime translations act on momentum space functions. To see this, consider the plane wave $e_p(x) = e^{ipx}$, regarded as a collection of functions on momentum space, labelled by x. Now take the translation $e_p(x + \varepsilon) = e^{ip(x+\varepsilon)} = e^{ip\varepsilon} e^{ipx} = e^{ip\varepsilon} e_p(x)$. Therefore, the translations act on momentum space as phase transformations.

The space translations with parameter $\boldsymbol{\varepsilon}$, we have

$$a_{\mathbf{p}} \mapsto e^{i\boldsymbol{\varepsilon}\mathbf{p}} a_{\mathbf{p}}, \quad b_{\mathbf{p}} \mapsto e^{i\boldsymbol{\varepsilon}\mathbf{p}} b_{\mathbf{p}} \tag{6.134}$$

$$a_{\mathbf{p}}^{\dagger} \mapsto e^{-i\boldsymbol{\varepsilon}\mathbf{p}} a_{\mathbf{p}}^{\dagger}, \quad b_{\mathbf{p}}^{\dagger} \mapsto e^{-i\boldsymbol{\varepsilon}\mathbf{p}} b_{\mathbf{p}}^{\dagger} \tag{6.135}$$

and the action (6.130) is clearly invariant; as for the time translations we take the phase factor to be $e^{\pm i\omega_{\mathbf{p}}\varepsilon}$ and the action is invariant again.

Next, the action is clearly rotational invariant, if we assume that $a_{\mathbf{p}}$, $b_{\mathbf{p}}$ are scalar functions of their arguments. It, therefore, remains to check the Lorentz invariance of the action. But since the action (6.130) has the form of the standard undeformed momentum space action, the transformation properties of the creation and annihilation 'operators' are just the standard ones $a_{\mathbf{p}} \mapsto U(\Lambda) a_{\mathbf{p}} U^{-1}(\Lambda) = a_{\mathbf{p}_\Lambda}$, where $\mathbf{p}_\Lambda$ is the spacial component of the Lorentz transformed four vector Λp. Indeed

$$
\begin{aligned}
U(\Lambda) S U^{-1}(\Lambda) &= \frac{1}{2} \int \frac{d^3 p}{2\omega_p} (p_\mu p^\mu + m^2) U(\Lambda) a_{\mathbf{p}}^{\dagger} a_{\mathbf{p}} U^{-1}(\Lambda) \\
&+ \frac{1}{2} \int \frac{d^3 p}{2\omega_p} (S(p)_\mu S(p)^\mu + m^2) U(\Lambda) b_{\mathbf{p}^*}^{\dagger} b_{\mathbf{p}^*} U^{-1}(\Lambda) \\
&= \frac{1}{2} \int \frac{d^3 p_\Lambda}{2\omega_{p_\Lambda}} ((\Lambda p)_\mu (\Lambda p)^\mu + m^2)) a_{\mathbf{p}_\Lambda}^{\dagger} a_{\mathbf{p}_\Lambda} \\
&+ \frac{1}{2} \int \frac{d^3 p_\Lambda}{2\omega_{p_\Lambda}} ((S(\Lambda p))_\mu (S(\Lambda p))^\mu + m^2) b_{\mathbf{p}_\Lambda}^{\dagger} b_{\mathbf{p}_\Lambda^*} = S
\end{aligned}
$$

This completes the proof of the Poincaré invariance of the action (6.130).

6.5 Symplectic Structure

The goal of this section is to derive the form of the creation/annihilation operators algebra in the case of the deformed complex scalar filed introduced in the next section. To this aim, we will use a very elegant covariant phase space method (see, e.g., [21] for a recent comprehensive review) which has the virtue of manifestly preserving the symmetries of the theory. To understand this method, let us consider the simplest possible theory, that of a bunch of non-relativistic particles. It is given by the action

$$S = \int_{\Sigma_{t_1}}^{\Sigma_{t_2}} dt \, \frac{1}{2} \sum_i m_i \dot{x}_i^2 - V(x_1, x_2, \ldots) \tag{6.136}$$

where for the future convenience, we denote the beginning and the end of trajectory by Σ_{t_1}, Σ_{t_2} even though these are just points. The variation of the action of the action can be written as

$$\delta S = -\int_{\Sigma_{t_1}}^{\Sigma_{t_2}} dt \sum_i \left(m_i \ddot{x}_i + \frac{\partial V(x)}{\partial x_i} \right) \delta x_i + \sum_i m_i \dot{x}_i \delta x_i \big|_{\Sigma_{t_1}}^{\Sigma_{t_2}} \qquad (6.137)$$

The bulk term is proportional to equations of motion and vanishes on-shell. We are left therefore with the boundary terms

$$\delta S = \sum_i m_i \dot{x}_i \delta x_i \big|_{\Sigma_{t_1}}^{\Sigma_{t_2}} = \sum_i \frac{\partial L}{\partial \dot{x}_i} \delta x_i \bigg|_{\Sigma_{t_1}}^{\Sigma_{t_2}} = \sum_i p_i \, \delta x_i \big|_{\Sigma_{t_1}}^{\Sigma_{t_2}} \qquad (6.138)$$

We define the presymplectic Liouville form (or Liouville form) as

$$\theta = -\sum_i p_i \, \delta x_i \big|_{\Sigma} \qquad (6.139)$$

The exterior differential of θ is called symplectic form and denoted ω

$$\omega = -\sum_i \delta p_i \wedge \delta x_i \big|_{\Sigma} = -\sum_{i,j} \omega^{ij} \delta p_i \wedge \delta x_i \bigg|_{\Sigma} \qquad (6.140)$$

The skew-symmetric bivector dual to the two-form ω defines the Poisson bracket

$$\{f(x, p), g(x, p)\} = \sum_{i,j} \omega_{ij} \frac{\partial f(x, p)}{\partial x_i} \frac{\partial f(x, p)}{\partial p_i} \qquad (6.141)$$

This construction can be readily extended to the case of field theory. There are two technical differences between the particle case and the field theory. First Σ_{t_1} and Σ_{t_2} are now three-dimensional spacelike submanifold of spacetime, which we can think of as constant time surfaces. Second the boundary terms at spacial infinity might be present, but for simplicity in what follows we neglect them. Finally, instead of the discrete index i labelling particles in field theory, we have to do with continuous position $\mathbf{x}$ labelling the fields. Apart from that the procedure in the case of field theory is a direct generalization of the one we discussed above, so we can start working it out.

Our starting point will be the variation of the action (6.82) which we computed in the preceding chapter (6.99), (6.104)

$$
\begin{aligned}
\delta S = \frac{1}{2} \int_{\mathbb{R}^4} d^4x \Bigg\{ & -\frac{\Delta_+}{\kappa} \left[(\partial_\mu^\dagger (\partial^\mu)^\dagger - m^2) \phi^\dagger \star \delta\phi \right] + \partial_A \left(\Pi_1^A \star \delta\phi \right) \\
& -\frac{\kappa}{\Delta_+} \left[\delta\phi^\dagger \star (\partial_\mu \partial^\mu - m^2)\phi \right] + \partial_A^\dagger \left(\delta\phi^\dagger \star \left(\Pi_1^A \right)^\dagger \right) \Bigg\} \\
+\frac{1}{2} \int_{\mathbb{R}^4} d^4x \Bigg\{ & - \left[\delta\phi \star (\partial_\mu^\dagger (\partial^\mu)^\dagger - m^2)\phi^\dagger \right] + \partial_A \left(\delta\phi \star \Pi_2^A \right) \\
& - \left[(\partial_\mu \partial^\mu - m^2)\phi \star \delta\phi^\dagger \right] + \partial_A^\dagger \left(\left(\Pi_2^A \right)^\dagger \star \delta\phi^\dagger \right) \Bigg\}
\end{aligned}
\qquad (6.142)
$$

where Π_1, Π_2 are defined in Equations (6.100)–(6.102) and (6.105)–(6.107). Disregarding the terms that vanish on-shell and the contribution form the boundary at spacial infinity for the presymplectic form we find

$$
\theta = -\frac{1}{2} \int_{\mathbb{R}^3} d^3x \left(\Pi_1^0 \star \delta\phi + \delta\phi^\dagger \star \left(\Pi_1^0 \right)^\dagger - \delta\phi \star \Pi_2^0 + \left(\Pi_2^0 \right)^\dagger \star \delta\phi^\dagger \right) \qquad (6.143)
$$

In deriving this equation, we use the fact (6.87) that

$$
\partial_0^\dagger = \partial_0 - i \Delta_+^{-1} \partial^2
$$

and notice that for the constant time boundary only the first term on the right hand side of this expression contributes. In the next step, we have to insert to the general expression (6.143) the mode expansion of the field and its adjoint (6.109), (6.110) alongside with the variations

$$
\begin{aligned}
\delta\phi(x) = & \int \frac{d^3p}{\sqrt{2\omega_p}} \left[1 + \frac{|p_+|^3}{\kappa^3} \right]^{-\frac{1}{2}} \delta a_{\mathbf{p}}\, e^{-i(\omega_p t - \mathbf{p}\mathbf{x})} \\
& + \int \frac{d^3p^*}{\sqrt{2|\omega_p^*|}} \left[1 + \frac{|p_+^*|^3}{\kappa^3} \right]^{-\frac{1}{2}} \delta b_{\mathbf{p}^*}^\dagger\, e^{i(S(\omega_p^*)t - S(\mathbf{p}^*)\mathbf{x})}
\end{aligned}
\qquad (6.144)
$$

and

$$
\begin{aligned}
\delta\phi^\dagger(x) = & \int \frac{d^3p}{\sqrt{2\omega_p}} \left[1 + \frac{|p_+|^3}{\kappa^3} \right]^{-\frac{1}{2}} \delta a_{\mathbf{p}}^\dagger\, e^{-i(S(\omega_p)t - S(\mathbf{p})\mathbf{x})} \\
& + \int \frac{d^3p^*}{\sqrt{2|\omega_p^*|}} \left[1 + \frac{|p_+^*|^3}{\kappa^3} \right]^{-\frac{1}{2}} \delta b_{\mathbf{p}^*}\, e^{i(\omega_p^* t - \mathbf{p}^*\mathbf{x})}
\end{aligned}
\qquad (6.145)
$$

The resulting computations are quite tedious, but the final expression is simple. Taking the external differential of the presymplectic form as we did above, we find the symplectic form of our deformed theory to be

$$\omega = \delta\theta = -\frac{i}{2}\int\frac{d^3p}{2\omega_p}\delta a_{\mathbf{p}} \wedge \delta a_{\mathbf{p}}^\dagger\,[\omega_p - S(\omega_p)]\frac{p_4}{\kappa} + \delta b_{\mathbf{p}*}^\dagger \wedge \delta b_{\mathbf{p}*}\,[S(\omega_p) - \omega_p]\frac{p_4}{\kappa}$$

(6.146)

From which we derive the following Poisson brackets

$$\left\{a_{\mathbf{p}}, a_{\mathbf{q}}^\dagger\right\} = i\,\frac{\kappa}{p_4}\,\frac{4\omega_p}{\omega_p - S(\omega_p)}\delta(\mathbf{p}-\mathbf{q})$$

(6.147)

$$\left\{b_{\mathbf{p}*}, b_{\mathbf{q}*}^\dagger\right\} = i\,\frac{\kappa}{p_4}\,\frac{4\omega_p}{\omega_p - S(\omega_p)}\delta(\mathbf{p}-\mathbf{q}).$$

(6.148)

Notice that the Poisson brackets for the 'particle' and 'antiparticle' creation/annihilation Fourier coefficients $a_{\mathbf{p}}$, $a_{\mathbf{q}}^\dagger$ and $b_{\mathbf{p}*}$, $b_{\mathbf{q}*}^\dagger$ are identical and very similar (up to the overall multiplicative factor) to their undeformed counterparts. Of course, when $\kappa \to \infty$ the difference between the undeformed creation/annihilation Poisson algebra and the deformed one (6.147), (6.148) disappears altogether.

6.6 Conserved Charges

By Noether theorem, with any global symmetry of the action, there are associated conserved charges. In this section, we construct the Noether charges related with translational symmetry, as we will see, rather unexpectedly, but consistently with the fact that the differential calculus is five-dimensional, we are going to find five independent charges of this kind.

Our starting point will be the again the action (6.82) which we recall here

$$S = \frac{1}{2}\int_{\mathbb{R}^4} d^4x\,\left[(\partial_\mu\phi)^\dagger \star \partial^\mu\phi + (\partial_\mu\phi)\star(\partial^\mu\phi)^\dagger + m^2(\phi^\dagger \star\phi + \phi\star\phi^\dagger)\right]$$

The variation of this action has been computed above and reads (6.99), (6.104)

$$\delta S = \frac{1}{2}\int_{\mathbb{R}_+^4} d^4x\,\left\{ -\frac{\Delta_+}{\kappa}\left[(\partial_\mu^\dagger(\partial^\mu)^\dagger - m^2)\phi^\dagger \star\delta\phi\right] + \partial_A\left(\Pi_1^A\star\delta\phi\right) \right.$$

$$\left. -\frac{\kappa}{\Delta_+}\left[\delta\phi^\dagger \star(\partial_\mu\partial^\mu - m^2)\phi\right] + \partial_A^\dagger\left(\delta\phi^\dagger\star\left(\Pi_1^A\right)^\dagger\right) \right\}$$

$$+\frac{1}{2}\int_{\mathbb{R}_+^4} d^4x\,\left\{ -\left[\delta\phi\star(\partial_\mu^\dagger(\partial^\mu)^\dagger - m^2)\phi^\dagger\right] + \partial_A\left(\delta\phi\star\Pi_2^A\right) \right.$$

$$\left. -\left[(\partial_\mu\partial^\mu - m^2)\phi\star\delta\phi^\dagger\right] + \partial_A^\dagger\left(\left(\Pi_2^A\right)^\dagger\star\delta\phi^\dagger\right) \right\}$$

where $\mathbb{R}^4_+$ indicates that we define the action on the half space with the boundary placed at $t = t_0$. The explicit expressions for the components of field momenta are given in (6.100)–(6.102) and (6.105)–(6.107).

The variation of the action is a sum of the terms that vanish when the field equations are imposed and total derivative terms. On-shell, therefore, the variation of the action reduces to the boundary term

$$\delta S_{on-shell} = \frac{1}{2} \int d^3x \, \Pi_1^0 \star \delta\phi + \delta\phi^\dagger \star \left(\Pi_1^0\right)^\dagger + \delta\phi \star \Pi_2^0 + \left(\Pi_2^0\right)^\dagger \star \delta\phi^\dagger \quad (6.149)$$

In the case of translations

$$\delta\phi = d\phi = \varepsilon^A \partial_A \phi \quad (6.150)$$

and we see that indeed we have to do with five independent symmetries, associated with five independent infinitesimal parameters ε^A. The five independent conserved Noether charges are then

$$\mathscr{P}_A = \frac{1}{2} \int d^3x \, T_1{}^0{}_A + T_2{}^0{}_A \quad (6.151)$$

where the relevant components of the energy–momentum tensor are

$$T_1{}^0{}_A = -\partial_A \Pi_1^0 \star \phi + \partial_A \phi^\dagger \star \Pi_1^{\dagger 0} \quad (6.152)$$

and

$$T_2{}^0{}_A = -\phi \star \partial_A \Pi_2^0 + \Pi_2^{\dagger 0} \star \partial_A \phi^\dagger \quad (6.153)$$

The charges (6.151) are conserved by construction, but we will see in a moment, after explicitly computing them, that they are indeed time independent.

Now we use the field decomposition (6.109), (6.110) to find the expression for conserved translational charges $\mathscr{P}_A$ in momentum space. After tedious computation, one finds that the time dependent terms cancel as they should and the conserved charges have the form

$$\mathscr{P}_0 = -\frac{1}{2} \int \frac{d^3p}{2\omega_p} a_\mathbf{p}^\dagger a_\mathbf{p} \, S(\omega_p)[\omega_p - S(\omega_p)]\frac{p_4}{\kappa} + b_{\mathbf{p}*} b_{\mathbf{p}*}^\dagger \, \omega_p[S(\omega_p) - \omega_p]\frac{p_4}{\kappa} \quad (6.154)$$

$$\mathscr{P}_i = \frac{1}{2} \int \frac{d^3p}{2\omega_p} a_\mathbf{p}^\dagger a_\mathbf{p} \, S(\mathbf{p})_i[\omega_p - S(\omega_p)]\frac{p_4}{\kappa} + b_{\mathbf{p}*} b_{\mathbf{p}*}^\dagger \, \mathbf{p}_i[S(\omega_p) - \omega_p]\frac{p_4}{\kappa} \quad (6.155)$$

$$\mathscr{P}_4 = -\frac{1}{2} \int \frac{d^3p}{2\omega_p} (p_4 - \kappa) \left\{ a_\mathbf{p}^\dagger a_\mathbf{p} \, [\omega_p - S(\omega_p)]\frac{p_4}{\kappa} + b_{\mathbf{p}*} b_{\mathbf{p}*}^\dagger \, \omega_p[S(\omega_p) - \omega_p]\frac{p_4}{\kappa} \right\}. \quad (6.156)$$

Notice that when $\kappa \to \infty$, $p_4 \to \kappa$, the charge $\mathscr{P}_4$ goes to zero, so it vanishes when the deformation disappears. We will discuss its physical meaning below in Sect. 7.3.

References

1. Freidel, L., Kowalski-Glikman, J., Nowak, S.: Field theory on kappa-Minkowski space revisited: Noether charges and breaking of Lorentz symmetry. Int. J. Mod. Phys. A **23**, 2687–2718 (2008). arXiv:0706.3658 [hep-th]
2. Arzano, M., Consoli, L.T.: Signal propagation on κ-Minkowski spacetime and nonlocal two-point functions. Phys. Rev. D **98**(10), 106018 (2018). arXiv:1808.02241 [hep-th]
3. Arzano, M., Bevilacqua, A., Kowalski-Glikman, J., Rosati, G., Unger, J.: κ-deformed complex fields and discrete symmetries. Phys. Rev. D **103**, 106015 (2021) arXiv:2011.09188 [hep-th]
4. Amelino-Camelia, G., Arzano, M.: Coproduct and star product in field theories on Lie algebra noncommutative space-times. Phys. Rev. D **65**, 084044 (2002). arXiv:hep-th/0105120 [hep-th]
5. Agostini, A., Amelino-Camelia, G., Arzano, M., Marciano, A., Tacchi, R.A.: Generalizing the Noether theorem for Hopf-algebra spacetime symmetries. Mod. Phys. Lett. A **22**, 1779–1786 (2007). arXiv:hep-th/0607221 [hep-th]
6. Arzano, M., Marciano, A.: Fock space, quantum fields and kappa-Poincare symmetries. Phys. Rev. D **76**, 125005 (2007). arXiv:0707.1329 [hep-th]
7. Arzano, M., Marciano, A.: Symplectic geometry and Noether charges for Hopf algebra space-time symmetries. Phys. Rev. D **75**, 081701 (2007). arXiv:hep-th/0701268 [hep-th]
8. Daszkiewicz, M., Lukierski, J., Woronowicz, M.: Towards quantum noncommutative kappa-deformed field theory. Phys. Rev. D **77**, 105007 (2008). arXiv:0708.1561 [hep-th]
9. Daszkiewicz, M., Lukierski, J., Woronowicz, M.: Kappa-deformed oscillators, the choice of star product and free kappa-deformed quantum fields. J. Phys. A **42**, 355201 (2009). arXiv:0807.1992 [hep-th]
10. Poulain, T., Wallet, J.C.: κ-Poincaré invariant quantum field theories with KMS weight. Phys. Rev. D **98**(2), 025002 (2018). arXiv:1801.02715 [hep-th]
11. Poulain, T., Wallet, J.C.: κ-Poincaré invariant orientable field theories at one-loop. JHEP **01**, 064 (2019). arXiv:1808.00350 [hep-th]
12. Kim, H.C., Lee, Y., Rim, C., Yee, J.H.: Scalar field theory in kappa-Minkowski spacetime from twist. J. Math. Phys. **50**, 102304 (2009). arXiv:0901.0049 [hep-th]
13. Govindarajan, T.R., Gupta, K.S., Harikumar, E., Meljanac, S., Meljanac, D.: Twisted statistics in kappa-Minkowski spacetime. Phys. Rev. D **77**, 105010 (2008). arXiv:0802.1576 [hep-th]
14. Govindarajan, T.R., Gupta, K.S., Harikumar, E., Meljanac, S., Meljanac, D.: Deformed oscillator algebras and QFT in kappa-Minkowski spacetime. Phys. Rev. D **80**, 025014 (2009). arXiv:0903.2355 [hep-th]
15. Meljanac, S., Samsarov, A., Trampetic, J., Wohlgenannt, M.: Noncommutative kappa-Minkowski phi4 theory: construction, properties and propagation. arXiv:1107.2369 [hep-th]
16. Mercati, F., Sergola, M.: Phys. Rev. D **98**(4), 045017 (2018). arXiv:1801.01765 [hep-th]
17. Majid, S., Ruegg, H.: Bicrossproduct structure of kappa Poincare group and noncommutative geometry. Phys. Lett. B **334**, 348 (1994). arXiv:hep-th/9405107
18. Sitarz, A.: Noncommutative differential calculus on the kappa Minkowski space. Phys. Lett. B **349**, 42 (1995). arXiv:hep-th/9409014
19. Bruno, N.R., Amelino-Camelia, G., Kowalski-Glikman, J.: Deformed boost transformations that saturate at the Planck scale. Phys. Lett. B **522**, 133–138 (2001). https://doi.org/10.1016/S0370-2693(01)01264-3, arXiv:hep-th/0107039 [hep-th]
20. Kowalski-Glikman, J., Walkus, A.: Star product and interacting fields on kappa-Minkowski space. Mod. Phys. Lett. A **24**, 2243 (2009). arXiv:0904.4036 [hep-th]
21. Harlow, D., Wu, J.Q.: Covariant phase space with boundaries. JHEP **10**, 146 (2020). arXiv:1906.08616 [hep-th]

Free Quantum Fields and Discrete Symmetries

7

In this final chapter, we continue investigating the theory of deformed scalar field uncovering its simplest quantum aspects.

As a first task, we perform a path integral quantization of the free scalar field to get an intuition of the physical effect of the κ-deformation on how external field perturbations are propagated by the non-commutative field.

We then discuss the basics of canonical quantization by computing the Poisson brackets between the mode functions $a_{\mathbf{q}}$, $a_{\mathbf{p}}^{\dagger}$, $b_{\mathbf{p}^*}$, $b_{\mathbf{q}^*}^{\dagger}$ which then will give rise to quantum commutators. We derive the expressions for conserved charges associated with Poincaté symmetry: energy, momentum, rotations, and boosts. Next, we discuss the subtle deformation of discrete symmetries $\mathscr{P}$, $\mathscr{T}$, $\mathscr{C}$, and in the final section, we present the possible observational traces of these deformations. The material presented in this chapter is based mostly on [1,2].

7.1 Free κ-Deformed Quantum Fields: The Feynman Propagator

7.1.1 The κ-Deformed Free Field Partition Function

The κ-Poincaré invariant action of a free massive complex scalar field is given by the formula (6.82), which we express here in the language of non-commutative fields $\hat{\phi}$ instead of the star product used before

$$S_{free}[\hat{\phi}, \hat{\phi}^{\dagger}] = -\frac{1}{2} \widehat{\int} \left[\left(\hat{\partial}_\mu \hat{\phi} \right)^{\dagger} \left(\hat{\partial}^\mu \hat{\phi} \right) + \left(\hat{\partial}_\mu \hat{\phi} \right) \left(\hat{\partial}^\mu \hat{\phi} \right)^{\dagger} + m^2 \left(\hat{\phi}^{\dagger} \hat{\phi} + \hat{\phi} \hat{\phi}^{\dagger} \right) \right].$$

$$(7.1)$$

© Springer-Verlag GmbH Germany, part of Springer Nature 2021
M. Arzano and J. Kowalski-Glikman, *Deformations of Spacetime Symmetries*,
Lecture Notes in Physics 986,
https://doi.org/10.1007/978-3-662-63097-6_7

From the action (7.1), making use of the coproduct properties of the $\hat{\partial}_\mu$'s, analogous to that studied in the previous section, (6.83)–(6.87) one obtains the following equation of motion (cf. (6.108))

$$(\hat{\partial}_\mu \hat{\partial}^\mu - m^2)\hat{\phi}(\hat{x}) = 0$$
$$((\hat{\partial}_\mu)^\dagger (\hat{\partial}^\mu)^\dagger - m^2)\hat{\phi}(\hat{x})^\dagger = 0 \tag{7.2}$$

which define identical mass-shells, reflecting the fact that, in the classical basis, the antipodes satisfy the relation

$$S(p_\mu)S(p)^\mu = p_\mu p^\mu . \tag{7.3}$$

In what follows, we will find it more convenient to use the following momentum space action that produces the same equations of motion

$$S_{free} = \frac{1}{2} \int \frac{d\mu(p)}{(2\pi)^4} \left[(p_\mu p^\mu + m^2) a(p)^\dagger a(p) + (S(p)_\mu S(p)^\mu + m^2) b(p)^\dagger b(p) \right] . \tag{7.4}$$

which is an off-shell counterpart of the momentum space action derived in the previous chapter, (6.130).

In what follows, for illustrative purposes we will consider only the 'particle sector' of the momentum space action, setting the 'antiparticle' Fourier components, $b, b^\dagger$ to zero. We leave it to the reader to repeat the steps presented below in the 'antiparticle' case.

The action in momentum space (7.4) can be used to write down the partition function of the theory. A partition function obtained from a momentum space action of a κ-deformed field was first used in [3]. A more recent use of the κ-deformed partition function, which implemented the non-trivial geometric features of the momentum space, has appeared in [4]. The partition function for the complex scalar field can be written as

$$\bar{Z}[J, J^\dagger] = \int \mathcal{D}[\phi]\mathcal{D}[\phi^\dagger] \, e^{iS_{free}[\hat{\phi}, \hat{\phi}^\dagger] + i \widehat{\int} \left[\hat{\phi}^\dagger \hat{J} + \hat{J}^\dagger \hat{\phi} \right]} , \tag{7.5}$$

where the action $S_{free}[\hat{\phi}, \hat{\phi}^\dagger]$ is the κ-Poincaré invariant action (7.1). We focus on the normalized partition function

$$Z[J, J^\dagger] = \frac{\bar{Z}[J, J^\dagger]}{\bar{Z}[0, 0]} . \tag{7.6}$$

In order to bring $Z[J, J^\dagger]$ into a well-suited expression for the manipulation needed to extract the Feynman propagator, we rewrite the partition function in momentum space. Indeed, since the momentum space is a commutative space, here it is possible to handle the functional calculus (which we will illustrate below) unambiguously. Keeping the 'particle' sector of the action (7.4), one obtains from (7.5)

$$Z[\tilde{J}, \tilde{J}^\dagger] = \frac{1}{Z[0,0]} \int \mathscr{D}[a]\mathscr{D}[a^\dagger] \, e^{i \int \frac{d\bar{\mu}(p)}{(2\pi)^4} \left[a^\dagger(p)(p_\mu p^\mu + m^2)a(p) + a^\dagger(p)\tilde{J}(p) + \tilde{J}^\dagger(p)a(p)\right]}.$$

(7.7)

The functional integration can now be carried out as an ordinary Gaussian integral, and after simple manipulations, one finds that

$$Z[\tilde{J}, \tilde{J}^\dagger] = \exp\left(i \int \frac{d\bar{\mu}(p)}{(2\pi)^4} \frac{\tilde{J}^\dagger(p)\tilde{J}(p)}{-p^2 - m^2 + i\varepsilon}\right).$$

(7.8)

In this last expression, we introduced the usual shift $m^2 \to m^2 - i\varepsilon$ to render the integral well defined.

In order to derive the Feynman propagator from the partition function $Z[\tilde{J}, \tilde{J}^*]$, we need an appropriate generalization of the functional derivatives to the deformed setting. In particular, one has to take into account the κ-deformed coproduct structure of the translation generators in (6.25), which leads to the following non-abelian addition laws for momenta:

$$(p \oplus q)_0 = p_0 q_+ + \frac{q_0}{p_+} + \frac{1}{\kappa}\frac{p_i q^i}{p_+}$$

(7.9)

$$(p \oplus q)_i = p_i q_+ + q_i.$$

(7.10)

This issue was first faced in [3], where, however, as recalled above, the explicit form of the momentum space integration measure was not taken into account. Nonetheless, for an explicit definition of the functional derivatives such information is needed. Indeed, an important ingredient in the construction of functional calculus is a notion of delta function on the space of momenta. We will consider a delta function compatible with the non-trivial momentum space measure $d\bar{\mu}(p)$ [5], i.e., such that

$$\int d\bar{\mu}(q) \, \delta(p,q) \, f(q) = f(p).$$

(7.11)

It can be easily checked that such delta function is given by

$$\delta(p,q) = \delta\big((\ominus p) \oplus q\big) = \frac{|p_4|}{\kappa} \delta(p-q),$$

(7.12)

and thus it is proportional to an ordinary delta function $\delta(p-q)$, and in particular, it is symmetric under the exchange of momenta in the argument $\delta\big((\ominus p) \oplus q\big) = \delta\big((\ominus q) \oplus p\big)$. Let us mention that the other possible choice of delta function $\delta\big(p \oplus (\ominus q)\big)$ would have been less natural since it carries and additional multiplicative factor, indeed from

$$\delta\big(p \oplus (\ominus q)\big) = |p_+|^3 \frac{|p_4|}{\kappa} \delta(p-q) = |p_+|^3 \delta\big((\ominus p) \oplus q\big),$$

(7.13)

it is easily seen that

$$\int d\bar{\mu}(q)\delta\big(p \oplus (\ominus q)\big) f(q) = |p_+|^3 f(p) \,. \tag{7.14}$$

Notice that the two delta functions $\delta\big((\ominus p) \oplus q\big)$ and $\delta\big(p \oplus (\ominus q)\big)$ are related by the antipode transformation $(p, q) \to (\ominus p, \ominus q)$, and thus the appearance of the factor $|p_+|^3$ is related to the Jacobian of the antipode map $\big|J\{\frac{\partial \ominus p}{\partial p}\}\big| = |p_+^{-1}|^3$.

With the choice of delta function (7.12), we can now proceed as in [3], though specializing the discussion to a complex field, and define the following κ-deformed functional derivatives:

$$\frac{\delta Z[\tilde{J}, \tilde{J}^\dagger]}{\delta \tilde{J}(q)} = \lim_{\varepsilon \to 0} \frac{1}{\varepsilon} \big\{ Z[\tilde{J}(p) + \varepsilon\delta((\ominus p) \oplus q), \tilde{J}^\dagger(p)] - Z[\tilde{J}, \tilde{J}^\dagger] \big\} \,,$$

$$\frac{\delta Z[\tilde{J}, \tilde{J}^\dagger]}{\delta \tilde{J}^\dagger(q)} = \lim_{\varepsilon \to 0} \frac{1}{\varepsilon} \big\{ Z[\tilde{J}(p), \tilde{J}^\dagger(p) + \varepsilon\delta((\ominus p) \oplus (\ominus q))] - Z[\tilde{J}, \tilde{J}^\dagger] \big\} \,, \tag{7.15}$$

which clearly reduce to the ordinary definitions in the limit $\kappa \to \infty$, given that $\ominus p \to -p$ and $p \oplus q \to p + q$. These will be employed in the next Section to obtain the Feynman propagator on κ-Minkowski non-commutative space.

7.1.2 The Feynman Propagator

We define the κ-deformed Feynman propagator in terms of the functional derivative of the partition function (7.8) with respect to incoming and outgoing source functions

$$i\tilde{\Delta}_F^\kappa(p, q) = \left(i\frac{\delta}{\delta \tilde{J}^\dagger(p)}\right)\left(-i\frac{\delta}{\delta \tilde{J}(q)}\right) Z[\tilde{J}, \tilde{J}^\dagger]\bigg|_{\tilde{J}, \tilde{J}^\dagger = 0}. \tag{7.16}$$

Taking into account the relations (7.15), one obtains

$$i\tilde{\Delta}_F^\kappa(p, q) = i(2\pi)^4 \frac{\delta((\ominus q) \oplus (\ominus p))}{-q^2 - m^2 + i\varepsilon} = i(2\pi)^4 \frac{|q_4|}{\kappa} \frac{\delta(q - S(p))}{-q^2 - m^2 + i\varepsilon}, \tag{7.17}$$

where in the last equality, we have expanded the delta function as in (7.12). Through the Fourier transform (6.50) it is then possible to obtain the free scalar Feynman propagator on κ-Minkowski non-commutative space

$$i\hat{\Delta}_F^\kappa(\hat{x}, \hat{y}) = i \int \frac{d\bar{\mu}(p)}{(2\pi)^4} \frac{\mathscr{W}^{-1}(e^{ipx})\big[\mathscr{W}^{-1}(e^{ipy})\big]^\dagger}{-p^2 - m^2 + i\varepsilon}. \tag{7.18}$$

where $\mathscr{W}$ is the Weyl map (6.62).

A point that deserves to be stressed, is that $i\hat{\Delta}_F^\kappa(\hat{x}, \hat{y})$ is not symmetric under exchange of its arguments. This property of the Feynman propagator (7.18) originates from the fact that, in the κ-deformed setting, the hermitian conjugate of a plane wave involves the antipode map $S(p)$ on its momentum. Such spacetime asymmetry of $i\hat{\Delta}_F^\kappa(\hat{x}, \hat{y})$ although it may seem puzzling, does not lead to an actual physical asymmetry of the κ-Minkowski field propagation, as we will show below. Nonetheless, the combination of non-commutative plane waves appearing in (7.18), which is the cause of this concern, makes sure that the Feynman propagator is a Green's function of the κ-Klein–Gordon equation (7.2), as we now show. As a first step, we define the non-commutative delta function $\hat{\delta}(\hat{x}, \hat{y})$ using the κ-Minkowski Fourier transform

$$\hat{f}(\hat{x}) = \int \frac{d\bar{\mu}(p)}{(2\pi)^4}\, \tilde{f}(p)\mathscr{W}^{-1}(e^{ipx}) = \int \frac{d\bar{\mu}(p)}{(2\pi)^4}\, \left(\widehat{\int_{\hat{y}} [\mathscr{W}^{-1}(e^{ipy})]^\dagger \hat{f}(\hat{y})}\right)\Omega_c(e^{ipx}).$$
(7.19)

By requiring

$$\hat{f}(\hat{x}) = \widehat{\int_{\hat{y}} \hat{\delta}(\hat{x}, \hat{y})\hat{f}(\hat{y})},$$
(7.20)

we are led to define the non-commutative δ-function[1]

$$\hat{\delta}(\hat{x}, \hat{y}) = \int \frac{d\bar{\mu}(p)}{(2\pi)^4}\mathscr{W}^{-1}(e^{ipx})[\mathscr{W}^{-1}(e^{ipy})]^\dagger.$$
(7.21)

Applying the κ-Klein–Gordon operator to the Feynman propagator (7.18), and taking into account the expression of the delta function (7.21), we thus get

$$(\hat{\partial}_\mu\hat{\partial}^\mu - m^2)\hat{\Delta}_F^\kappa(\hat{x}, \hat{y}) = -\hat{\delta}(\hat{x}, \hat{y}),$$
(7.22)

which shows that $\hat{\Delta}_F^\kappa(\hat{x}, \hat{y})$ is a (non-commutative) Green's function for this operator.

We now get back to the issue of the spacetime asymmetry of the κ-deformed propagator raised above and study how $\hat{\Delta}_F^\kappa(\hat{x}, \hat{y})$ propagates the field in the presence of a perturbation generated by an external source $\hat{J}(\hat{x})$. This exercise will also provide significant insight on the physical properties of the κ-Minkowski Feynman propagator. Given the κ-Klein–Gordon equation in the presence of a source

$$(\hat{\partial}_\mu\hat{\partial}^\mu - m^2)\hat{\phi}(\hat{x}) = \hat{J}(\hat{x}),$$
(7.23)

---

[1] The non-commutative delta function $\hat{\delta}(\hat{x}, \hat{y})$ also satisfies:

$$\widehat{\int_{\hat{x},\hat{y}} \hat{\delta}(\hat{x}, \hat{y})} = 1.$$

using the Eqs. (7.20) and (7.22), we can write down the following solution:

$$\hat{\phi}(\hat{x}) = - \widehat{\int_{\hat{y}} \hat{\Delta}_F^{\kappa}(\hat{x}, \hat{y}) \hat{J}(\hat{y})} \, . \tag{7.24}$$

In contrast with the standard commutative case, the κ-Minkowski integral (7.24) involves the product of two non-commutative functions and requires some additional care. As we saw in the previous chapter, such integral can be defined via the ordinary Lebesgue integral introducing a non-commutative $\star$-multiplication between the fields. Indeed, writing the source $\hat{J}(\hat{y})$ as a Fourier integral

$$\hat{J}(\hat{y}) = \int \frac{d\bar{\mu}(p)}{(2\pi)^4} \, \tilde{J}(p) \, \Omega_c(e^{ipy}) \, , \tag{7.25}$$

we see that (7.24) involves an integral of the form

$$\widehat{\int_{\hat{y}} [\mathcal{W}^{-1}(e^{ipy})]^{\dagger} \mathcal{W}^{-1}(e^{iqy})} \, . \tag{7.26}$$

Making use of the Weyl map $\mathcal{W}$ and its associated star product, can be expressed as

$$\widehat{\int_{\hat{y}} [\mathcal{W}^{-1}(e^{ipy})]^{\dagger} \mathcal{W}^{-1}(e^{iqy})} = \int d^4 y \, \left(e^{iS(p)_{\mu} y^{\mu}}\right) \star \left(e^{iq_{\mu} y^{\mu}}\right) =$$
$$= \int d^4 y \, e^{-ip_{\mu} y^{\mu}} \sqrt{1 + \Box/\kappa^2} \, e^{iq_{\mu} y^{\mu}} \, . \tag{7.27}$$

We have, therefore, that the propagation law (7.24) can be expressed in terms of commutative fields as

$$\phi(x) = - \int d^4 y \, \Delta_F^{\kappa}(x - y) \sqrt{1 + \Box/\kappa^2} \, J(y) \, , \tag{7.28}$$

where $\phi(x) = \mathcal{W}(\hat{\phi}(\hat{x}))$, $J(y) = \mathcal{W}(\hat{J}(\hat{y}))$ and we defined the κ-deformed Feynman propagator on *commutative* Minkowski space

$$i\Delta_F^{\kappa}(x - y) = i \int \frac{d\bar{\mu}(p)}{(2\pi)^4} \frac{e^{ip(x-y)}}{-p^2 - m^2 + i\varepsilon} \, . \tag{7.29}$$

From (7.28), we see that the spacetime asymmetry of the κ-Minkowski Feynman propagator (7.18) does not affect the actual propagation of the field. Indeed, such asymmetry is cancelled by the star product of (7.27), leading to a field propagation governed by the spacetime symmetric κ-deformed Feynman propagator (7.29).

Let us also notice that in the propagation law (7.28), we can make the star product term $\sqrt{1 + \Box/\kappa^2}$ act either on the source $J(y)$ or, equivalently,[2] on the κ-deformed propagator $\Delta_F^\kappa(x - y)$.

Acting on $J(y)$ and noticing that in momentum space the term $\sqrt{1 + \Box/\kappa^2}$ is equal to $|p_4|/\kappa$, so that it cancels the same factor in the non-trivial integration measure $d\tilde{\mu}(p)$, one obtains

$$\phi(x) = - \int d^4y \, \Delta_F^\kappa(x - y) J_{cl}(y) , \tag{7.30}$$

where $J_{cl}(y)$ is a *classical* source

$$J_{cl}(y) = \int \frac{d^4p}{(2\pi)^4} \, \tilde{J}(p) \, e^{ipy} , \tag{7.31}$$

i.e., just an ordinary commutative function which, in particular, can describe a sharply localized source (e.g., a Dirac delta function).

Acting instead with the star product term $\sqrt{1 + \Box/\kappa^2}$ in (7.28) on $\Delta_F^\kappa(x - y)$, one gets

$$\phi(x) = - \int d^4y \, \Delta_F(x - y) J(y) , \tag{7.32}$$

where $i\,\Delta_F(x - y)$ is the undeformed free scalar Feynman propagator

$$i\,\Delta_F(x - y) = i \int \frac{d^4p}{(2\pi)^4} \, \frac{e^{ip(x-y)}}{-p^2 - m^2 + i\varepsilon} , \tag{7.33}$$

and where, in this case, the source function $J(y)$ can not describe a point-like source due to the presence of the κ-deformed integration measure $d\tilde{\mu}(p) = d^4p \, \theta(\kappa^2 - p^2)\kappa/|p_4|$ in its Fourier expansion (7.25). The source function $J(y)$ can indeed be seen as a *smeared* version of the classical source (7.31). For instance, considering a classical source sharply localized in space $J_{cl}(y) = \delta(\mathbf{y})$, for which $\tilde{J}(p) = 2\pi \delta(p_0)$, the source $J(y)$ takes the form

$$J(y) = 2 \int_0^\kappa \frac{dp \, p}{(2\pi)^2 \sqrt{1 - p^2/\kappa^2}} \, \frac{\sin(p|\mathbf{y}|)}{|\mathbf{y}|} = \frac{\kappa^2}{4\pi} \, \frac{J_1(\kappa|\mathbf{y}|)}{|\mathbf{y}|} , \tag{7.34}$$

where J_1 is the Bessel function of the first kind (Fig. 7.1).

Summarizing, we have the following two pictures for the propagation of the non-commutative κ-scalar field in terms of commuting fields

---

[2] Given the propagation law $\phi(x) = - \int d^4y \, \Delta_F^\kappa(x - y) \, \sqrt{1 + \Box/\kappa^2} \, J(y)$ we can take the formal series expansion in powers of the d'Alembertian for the star product term $\phi(x) = - \int d^4y \, \sum_{n=0}^\infty a_n \Delta_F^\kappa(x - y) \, \Box^n J(y)$ which, after integrating by parts becomes $\phi(x) = - \int d^4y \, \sum_{n=0}^\infty a_n \Box^n \Delta_F^\kappa(x - y) \, J(y)$.

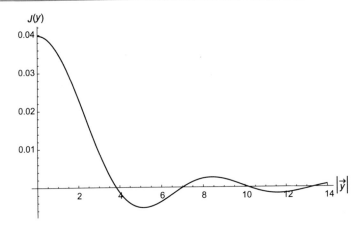

Fig. 7.1 Smeared version $J(y)$ of a classical source sharply localized in space $J_{cl}(y) = \delta(\mathbf{y})$. Here we have set the deformation parameter $\kappa = 1$

1. Given a perturbation generated by a classical, and virtually sharply localized source $J_{cl}(y)$, the field responds by propagating through the κ-deformed Feynman propagator $i\Delta^{\kappa}_F(x - y)$.
2. Given a perturbation generated by a κ-deformed source $J(y)$, the field responds by propagating through the standard Feynman propagator $i\Delta_F(x - y)$.

This result provides a concrete realization of the picture qualitatively outlined in [4]. There it was showed that for κ-deformed fields, the Yukawa potential between two static point sources does not diverge in the short-distance limit and that this feature could be interpreted in terms of point-like sources being effectively smoothed out by the UV features of the κ-deformation. This is precisely what is realized in the propagation picture (7.32) outlined above.

7.2   Creation/Annihilation Operators Algebra

In quantum theory, the Poisson brackets (6.147), (6.148) become commutators of quantum operators $\hat{a}_{\mathbf{p}}$, $\hat{a}^{\dagger}_{\mathbf{q}}$ and $\hat{b}_{\mathbf{p}^*}$, $\hat{b}^{\dagger}_{\mathbf{q}^*}$

$$\left[\hat{a}_{\mathbf{p}}, \hat{a}^{\dagger}_{\mathbf{q}}\right] = \frac{\kappa}{p_4} \frac{4\omega_p}{\omega_p - S(\omega_p)} \delta(\mathbf{p} - \mathbf{q}) \tag{7.35}$$

$$\left[\hat{b}_{\mathbf{p}^*}, \hat{b}^{\dagger}_{\mathbf{q}^*}\right] = \frac{\kappa}{p_4} \frac{4\omega_p}{\omega_p - S(\omega_p)} \delta(\mathbf{p} - \mathbf{q}). \tag{7.36}$$

We define the vacuum state $|0\rangle$ which is being annihilated by all annihilation operators

$$\hat{a}_{\mathbf{p}} |0\rangle = \hat{b}_{\mathbf{p}} |0\rangle = 0 \tag{7.37}$$

In the next step, we can define the one-particle and one-antiparticle states

$$|\mathbf{p}\rangle_a \equiv a_{\mathbf{p}}^\dagger |0\rangle \qquad (7.38)$$

$$|\mathbf{p}\rangle_b \equiv b_{\mathbf{p}*}^\dagger |0\rangle \qquad (7.39)$$

Acting with a creation operator on the one-(anti)particle state, we would produce a two-particle state, then the three-particle state, and so on. In this way, in the standard theory, one constructs the Fock space, the space of all possible multi-particles states we formally introduced in Sect. 5.1. In the κ-deformed case, the construction of Fock space is far more difficult because one must take into account the coproduct structure of κ-Poincaré algebra. Let us make a brief digression to illustrate this point, since this is one of the main unresolved issues in the development of a consistent κ-deformed quantum field theory. As we discussed in Sect. 5.1, the coproduct of a generator of the Poincaré algebra simply tells us how they act on tensor product states. For elementary relativistic particle, however, there is a subtlety: they are indistinguishable and thus a simple tensor product of n one-particle states cannot represent the actual quantum state of a n-particles system. One must, instead, consider *symmetrized* or *antisymmetrized* tensor product states when considering bosons and fermions, respectively. Let us focus for simplicity on an *undeformed* two-particle bosonic system. The state in ordinary Fock space representing such state is

$$|\mathbf{p}_1\mathbf{p}_2\rangle = \frac{1}{\sqrt{2}} \left(|\mathbf{p}_1\rangle \otimes |\mathbf{p}_2\rangle + |\mathbf{p}_2\rangle \otimes |\mathbf{p}_1\rangle \right), \qquad (7.40)$$

where the normalization factor ensures that

$$\langle \mathbf{p}_1\mathbf{p}_2 | \mathbf{p}_1\mathbf{p}_2\rangle = 1 \qquad (7.41)$$

Now the action of the generator of space translations on the state (7.40) is expressed in terms of the trivial coproduct (5.8)

$$\Delta P_i = P_i \otimes 1 + 1 \otimes P_i \qquad (7.42)$$

telling us that the total momentum of a two-particle state is simply the sum of the momenta of the individual particles

$$\Delta P_i \, |\mathbf{p}_1\mathbf{p}_2\rangle = (\mathbf{p}_1 + \mathbf{p}_2)|\mathbf{p}_1\mathbf{p}_2\rangle . \qquad (7.43)$$

Now, one of the characterizing features of the κ-Poincaré Hopf algebra is that the coproduct for spatial translation generators is non-commutative and thus, unlike in the deformed case, the two states $|\mathbf{p}_1\rangle \otimes |\mathbf{p}_2\rangle$ and $|\mathbf{p}_2\rangle \otimes |\mathbf{p}_1\rangle$ have *different momentum eigenvalues*. This implies that assuming an ordinary symmetrization in the deformed case would lead to a multi-particle states which *are not* eigenstates of the generators of spatial translations, and thus do not have a well defined spatial momentum. A way out of this unpleasant situation was first proposed in [6] (see also [7] for an earlier discussion in the context of κ-deformed oscillators) and then refined in [8] and consisted of replacing an ordinary symmetrization with a deformed symmetrization, e.g., for a two-particle state to consider the following combination of tensor products

$$|p_1 p_2\rangle_\kappa = \frac{1}{\sqrt{2}} \left[|\mathbf{p}_1\rangle \otimes |\mathbf{p}_2\rangle + |(1 - \varepsilon_1)\mathbf{p}_2\rangle \otimes |(1 - \varepsilon_2)^{-1}\mathbf{p}_1\rangle \right]$$

$$|p_2 p_1\rangle_\kappa = \frac{1}{\sqrt{2}} \left[|\mathbf{p}_2\rangle \otimes |\mathbf{p}_1\rangle + |(1 - \varepsilon_2)\mathbf{p}_1\rangle \otimes |(1 - \varepsilon_1)^{-1}\mathbf{p}_2\rangle \right] ,$$

where $1 - \varepsilon_i = 1 - \frac{|\mathbf{p}_i|}{\kappa} = e^{-\omega(\mathbf{p}_i)/\kappa}$. It can be easily seen that these states are eigenstates of the generators of κ-deformed translations in the bicrossproduct basis with eigenvalues $\mathbf{p}_1 \oplus \mathbf{p}_2$ and $\mathbf{p}_2 \oplus \mathbf{p}_1$ (where $\oplus$ stands for the non-abelian composition of momenta (6.17)), respectively. This recipe can be extended to general n-particle states, thus leading to the construction of a full κ-deformed Fock space. There are, however, two drawbacks of such construction. The first is that it is quite difficult to formulate an algebra of creation and annihilation operators which acting repeatedly on the vacuum would produce the multi-particle states of the κ-deformed Fock space [8]. The second, and more severe problem, is that the κ-deformed multi-particle states constructed in this way are covariant under deformed boosts *only at leading order in* $1/\kappa$ [8]. A possible solution to this problem was proposed in [9] with the introduction of a coassociator and a quasibialgebra structure for κ-Poincaré, thus leaving the standard Hopf algebra realm. As it stands, however, the problem of multi-particle states is still considered an open one.

Now we are in a position to define quantum operators associated with translational symmetries. They are given by operator analogues of the conserved charges constructed in Sect. 6.6.

In quantum theory, the charge $\mathscr{P}_0$ becomes the Hamiltonian operator

$$\hat{H} = -\frac{1}{2} \int \frac{d^3 p}{2\omega_p} \hat{a}_{\mathbf{p}}^\dagger \hat{a}_{\mathbf{p}} \, S(\omega_p)[\omega_p - S(\omega_p)]\frac{p4}{\kappa} + \hat{b}_{\mathbf{p}^*}^\dagger \hat{b}_{\mathbf{p}^*} \, \omega_p [S(\omega_p) - \omega_p]\frac{p4}{\kappa}$$

$$(7.44)$$

In writing the Hamiltonian operator, we normally ordered the creation and annihilation operators, so as to make the energy of the vacuum equal zero. We will normal order these operators also below to make the vacuum chargeless.

The spacial components of $\mathscr{P}$ become the components of field momentum operator

$$\hat{\mathbf{P}} = \frac{1}{2} \int \frac{d^3 p}{2\omega_p} \hat{a}_{\mathbf{p}}^\dagger \hat{a}_{\mathbf{p}} \, \mathbf{S}(\mathbf{p})[\omega_p - S(\omega_p)]\frac{p4}{\kappa} + \hat{b}_{\mathbf{p}^*}^\dagger \hat{b}_{\mathbf{p}^*} \, \mathbf{p}[S(\omega_p) - \omega_p]\frac{p4}{\kappa} \quad (7.45)$$

Finally, we have the 'fifth-momentum' operator

$$\hat{P}_4 = -\frac{1}{2} \int \frac{d^3 p}{2\omega_p} (p_4 - \kappa) \left\{ \hat{a}_{\mathbf{p}}^\dagger \hat{a}_{\mathbf{p}} [\omega_p - S(\omega_p)] \frac{p_4}{\kappa} + \hat{b}_{\mathbf{p}^*}^\dagger \hat{b}_{\mathbf{p}^*} [S(\omega_p) - \omega_p] \frac{p_4}{\kappa} \right\}.$$

$$(7.46)$$

whose interpretation we will discuss in Sect. 7.3.

7.3   Discrete Symmetries

Continuous global symmetries, like Poincaré symmetry, are not the only symmetries relevant for quantum field theory. Very important are also three discrete symmeries: parity $\mathscr{P}$, time reversal $\mathscr{T}$, and charge conjugation $\mathscr{C}$. Although the combination of two $\mathscr{C}\mathscr{P}$ can be broken, it is believed that the product of three of them $\mathscr{C}\mathscr{P}\mathscr{T}$ is an exact symmetry of any undeformed quantum field theory. In fact, there are very close relations between $\mathscr{C}\mathscr{P}\mathscr{T}$ and Poincaré symmetries: in undeformed theories under mild assumptions $\mathscr{C}\mathscr{P}\mathscr{T}$ violation implies Lorentz violation [10]. For thorough discussion of discrete symmetries and their experimental confirmation see [11]. In this section, we will first present each of the discrete symmetries in the undeformed case, for the field

$$\hat{\phi}(t, \mathbf{x})_{\kappa=0} = \int \frac{d^3 p}{\sqrt{2\omega_p}} \hat{a}_{\mathbf{p}} e^{-i(\omega_p t - i\mathbf{p}\mathbf{x})} + \hat{b}_{\mathbf{p}}^\dagger e^{i(\omega_p t - \mathbf{p}\mathbf{x})} \qquad (7.47)$$

and then generalize it to the case of the deformed fields (6.109) and (6.110).

7.3.1   Parity $\mathscr{P}$

The parity operator $\mathscr{P}$ acting on spacetime coordinates results in inversion $x = (t, \mathbf{x}) \rightarrow x' = (t, -\mathbf{x})$. In the case of an undeformed complex scalar quantum field, the action of parity operator results in the field at the point with inverted space coordinate

$$\mathscr{P}\hat{\phi}(t, \mathbf{x})_{\kappa=0}\mathscr{P}^{-1} = \int \frac{d^3 p}{\sqrt{2\omega_p}} \mathscr{P}\hat{a}_{\mathbf{p}}\mathscr{P}^{-1} e^{-i(\omega_p t - \mathbf{p}\mathbf{x})} + \mathscr{P}\hat{b}_{\mathbf{p}}^\dagger \mathscr{P}^{-1} e^{i(\omega_p t - \mathbf{p}\mathbf{x})}$$

$$(7.48)$$

On the other hand, we have

$$\phi(t, -\mathbf{x})_{\kappa=0} = \int \frac{d^3 p}{\sqrt{2\omega_p}} \hat{a}_{\mathbf{p}} e^{-i(\omega_p t + i\mathbf{p}\mathbf{x})} + \hat{b}_{\mathbf{p}}^\dagger e^{i(\omega_p t + \mathbf{p}\mathbf{x})} \qquad (7.49)$$

By making the momentum inversions $\mathbf{p} \mapsto -\mathbf{p}$ in the second expression and noticing that the integration measure is invariant under inversion, we see that for the
creation/annihilation operators[3]

$$\mathscr{P}\hat{a}_{\mathbf{p}}\mathscr{P}^{-1} = \hat{a}_{-\mathbf{p}}, \quad \mathscr{P}\hat{b}_{\mathbf{p}}\mathscr{P}^{-1} = \hat{b}_{-\mathbf{p}} \tag{7.50}$$

In the deformed case, we first convince ourselves that the spacetime transformation $\hat{\mathbf{x}} \mapsto -\hat{\mathbf{x}}$ leaves the invariant defining commutator (6.1)

$$[\hat{x}^0, \hat{x}^i] = \frac{i}{\kappa}\hat{x}^i, \quad [\hat{x}^i, \hat{x}^j] = 0, \quad i, j = 1, 2, 3.$$

Therefore, parity is compatible with the form of κ-Minkowski non-commutativity.
Then we notice that in the deformed case, nothing really changes as compared to
the undeformed case. The only difference is that instead of the change of sign in the
exponent in the second term in (7.47), we now have to do with the antipode. But the
antipode also changes sign under momentum inversion

$$S(p_0, -\mathbf{p})_0 = S(p_0, \mathbf{p})_0, \quad S(p_0, -\mathbf{p})_i = -S(p_0, \mathbf{p})_i$$

so that in the deformed case, parity acts on creation/annihilation operators as in the
undeformed case

$$\mathscr{P}\hat{a}_{\mathbf{p}}\mathscr{P}^{-1} = a_{-\mathbf{p}}, \quad \mathscr{P}\hat{b}_{\mathbf{p}^*}\mathscr{P}^{-1} = \hat{b}_{-\mathbf{p}^*} \tag{7.51}$$

and

$$\mathscr{P}\hat{a}_{\mathbf{p}}^\dagger\mathscr{P}^{-1} = \hat{a}_{-\mathbf{p}}^\dagger, \quad \mathscr{P}\hat{b}_{\mathbf{p}^*}^\dagger\mathscr{P}^{-1} = \hat{b}_{-\mathbf{p}^*}^\dagger \tag{7.52}$$

7.3.2 Time Reversal $\mathscr{T}$

Next, we consider the time reversal $\mathscr{T}$, which changes the direction of time $x = (t, \mathbf{x}) \to x' = (-t, \mathbf{x})$ and

$$\mathscr{T}\phi(t, \mathbf{x})\mathscr{T}^{-1} = \phi(-t, \mathbf{x}) \tag{7.53}$$

It should be remembered that the operator $\mathscr{T}$ is anti-hermitian $\mathscr{T}i\mathscr{T}^{-1} = -i$, and
we have

$$\mathscr{T}\hat{\phi}(t, \mathbf{x})_{\kappa=0}\mathscr{T}^{-1} = \int \frac{d^3p}{\sqrt{2\omega_p}} \mathscr{T}\hat{a}_{\mathbf{p}}\, e^{-i(\omega_p t - i\mathbf{p}\mathbf{x})}\, \mathscr{T}^{-1} + \mathscr{T}\hat{b}_{\mathbf{p}}^\dagger\, e^{i(\omega_p t - \mathbf{p}\mathbf{x})}\, \mathscr{T}^{-1}$$

$$= \int \frac{d^3p}{\sqrt{2\omega_p}} \mathscr{T}\hat{a}_{\mathbf{p}}\mathscr{T}^{-1}\, e^{i(\omega_p t - i\mathbf{p}\mathbf{x})} + \mathscr{T}\hat{b}_{\mathbf{p}}^\dagger\mathscr{T}^{-1}\, e^{-i(\omega_p t - \mathbf{p}\mathbf{x})} \tag{7.54}$$

---

[3]Here and below, we ignore a possible phase factor that may be present in the definition.

We have also

$$\hat{\phi}(-t, \mathbf{x})_{\kappa=0} = \int \frac{d^3 p}{\sqrt{2\omega_p}} \, \hat{a}_{\mathbf{p}} \, e^{-i(-\omega_p t - i\mathbf{px})} + \hat{b}_{\mathbf{p}}^\dagger \, e^{i(-\omega_p t - \mathbf{px})} \tag{7.55}$$

We find that

$$\mathscr{T}\hat{a}_{\mathbf{p}}\mathscr{T}^{-1} = \hat{a}_{-\mathbf{p}}, \quad \mathscr{T}\hat{b}_{\mathbf{p}}\mathscr{T}^{-1} = \hat{b}_{-\mathbf{p}}. \tag{7.56}$$

Let us now discuss the deformed case. We start noticing that as a consequence of anti-hermiticity of $\mathscr{T}$ the defining algebra (6.1) is again invariant, since $\mathscr{T}$ changes sign of $\hat{x}^0$ and the sign in front of i on the right hand side, so that we see that κ-Minkowski space is both parity and time reversal invariant. Turning to the deformed fields, we again see that the reasoning that we presented for the undeformed case above can be verbatim repeated in the case of time reversal as well and we end up with

$$\mathscr{T}\hat{a}_{\mathbf{p}}\mathscr{T}^{-1} = \hat{a}_{-\mathbf{p}}, \quad \mathscr{T}\hat{b}_{\mathbf{p}^*}\mathscr{T}^{-1} = \hat{b}_{-\mathbf{p}^*} \tag{7.57}$$

and

$$\mathscr{T}\hat{a}_{\mathbf{p}}^\dagger\mathscr{T}^{-1} = \hat{a}_{-\mathbf{p}}^\dagger, \quad \mathscr{T}\hat{b}_{\mathbf{p}^*}^\dagger\mathscr{T}^{-1} = \hat{b}_{-\mathbf{p}^*}^\dagger \tag{7.58}$$

7.3.3   Charge Conjugation $\mathscr{C}$

The symmetry that exchanges particles with antiparticles does not have any spacetime counterparts, and since it changes the charges carried by particles and antiparticles, it is called charge conjugation. In the undeformed case, the charge conjugation operator $\mathscr{C}$ acting on the field produces its conjugation

$$\mathscr{C}\hat{\phi}(t, \mathbf{x})_{\kappa=0}\mathscr{C}^{-1} = \hat{\phi}^\dagger(t, \mathbf{x})_{\kappa=0} \tag{7.59}$$

therefore

$$\mathscr{C}\hat{\phi}(t, \mathbf{x})_{\kappa=0}\mathscr{C}^{-1} = \int \frac{d^3 p}{\sqrt{2\omega_p}} \, \mathscr{C}\hat{a}_{\mathbf{p}}\mathscr{C}^{-1} \, e^{-i(\omega_p t - i\mathbf{px})} + \mathscr{C}\hat{b}_{\mathbf{p}}^\dagger\mathscr{C}^{-1} \, e^{i(\omega_p t - \mathbf{px})} =$$

$$\hat{\phi}^\dagger(t, \mathbf{x})_{\kappa=0} = \int \frac{d^3 p}{\sqrt{2\omega_p}} \, \hat{a}_{\mathbf{p}}^\dagger \, e^{i(\omega_p t - i\mathbf{px})} + \hat{b}_{\mathbf{p}} \, e^{-i(\omega_p t - \mathbf{px})} \tag{7.60}$$

and we have

$$\mathscr{C} a_{\mathbf{p}} \mathscr{C}^{-1} = b_{\mathbf{p}} \tag{7.61}$$

Let us now consider the deformed field. Consider the positive energy component $\phi_{(+)}$ first

$$\mathscr{C}\hat{\phi}_{(+)}(t, \mathbf{x})\mathscr{C}^{-1} = \int \frac{d^3 p}{\sqrt{2\omega_p}} \left[1 + \frac{|p_+|^3}{\kappa^3} \right]^{-\frac{1}{2}} \mathscr{C}\hat{a}_{\mathbf{p}}\mathscr{C}^{-1} \, e^{-i(\omega_p t - i\mathbf{px})} \tag{7.62}$$

On the other hand, we have, for the positive energy component of the adjoint field
$\hat{\phi}^\dagger_{(-)}(x)$

$$\hat{\phi}^\dagger_{(-)}(x) = \int \frac{d^3 p^*}{\sqrt{2|\omega^*_p|}} \left[1 + \frac{|p_+|^3}{\kappa^3}\right]^{-\frac{1}{2}} \hat{b}_{\mathbf{p}^*} \, e^{i(\omega^*_p t - \mathbf{p}^* \mathbf{x})} \tag{7.63}$$

from which we deduce that

$$\mathscr{C} \hat{a}_{\mathbf{p}} \mathscr{C}^{-1} = \hat{b}_{\mathbf{p}^*} \tag{7.64}$$

Analogously, for the positive energy component $\phi_{(-)}$

$$\mathscr{C} \hat{\phi}_{(-)}(x) \mathscr{C}^{-1} = \int \frac{d^3 p^*}{\sqrt{2|\omega^*_p|}} \left[1 + \frac{|p_+|^3}{\kappa^3}\right]^{-\frac{1}{2}} \mathscr{C} \hat{b}^\dagger_{\mathbf{p}^*} \mathscr{C}^{-1} \, e^{i(S(\omega^*_p)t - S(\mathbf{p}^*)\mathbf{x})} \tag{7.65}$$

and the negative energy component of the adjoint field $\hat{\phi}^\dagger_{(+)}(x)$

$$\hat{\phi}^\dagger_{(+)}(x) = \int \frac{d^3 p}{\sqrt{2\omega_p}} \left[1 + \frac{|p_+|^3}{\kappa^3}\right]^{-\frac{1}{2}} \hat{a}^\dagger_{\mathbf{p}} \, e^{-i(S(\omega_p)t - S(\mathbf{p})\mathbf{x})} \tag{7.66}$$

So that

$$\mathscr{C} \hat{b}^\dagger_{\mathbf{p}^*} \mathscr{C}^{-1} = \hat{a}^\dagger_{\mathbf{p}} \tag{7.67}$$

It should be stressed that these simple transformation rules of the field $\hat{\phi}$ with respect to charge conjugation is a result of the use of the second (starred) copy of momentum space and of the particular arrangement of the components $\hat{\phi}_{(\pm)}(x)$ and $\hat{\phi}^\dagger_{(\pm)}(x)$. In particular, the field constructed in [12] and many other papers on this topic does not transform nicely under charge conjugation. Also, it should be added that the deformed action of discrete symmetries $\mathscr{P}$, $\mathscr{T}$, and $\mathscr{C}$ leads to the form of the $\mathscr{C}\mathscr{P}\mathscr{T}$ operator Θ anticipated in the paper [13], where some phenomenological implications of deformed $\mathscr{C}\mathscr{P}\mathscr{T}$ invariance were considered.

7.4 Properties of One-Particle States

As we mentioned in Sect. 7.3, we still do not fully understand the construction of Fock space in the κ-deformed case. However, the single-particle and/or single antiparticle sectors can be analyzed with the tools that we already have developed.

We start with the vacuum state $|0\rangle$ (7.37) that is annihilated by all annihilation operators and then we define the one-particle and one-antiparticle states

$$|\mathbf{p}\rangle_a \equiv a^\dagger_{\mathbf{p}} |0\rangle , \tag{7.68}$$

$$|\mathbf{p}\rangle_b \equiv b^\dagger_{\mathbf{p}^*} |0\rangle . \tag{7.69}$$

To see that the names 'one-particle' and 'one-antiparticle' state to denote the states above is indeed justified consider the action of the charge conjugation operator $\mathscr{C}$ that we constructed in the preceding Sect. 7.3. We have

$$\mathscr{C}\,|\mathbf{p}\rangle_b = \mathscr{C}\hat{b}_\mathbf{p}^\dagger\mathscr{C}^{-1}\,\mathscr{C}\,|0\rangle = \mathscr{C}\hat{b}_\mathbf{p}^\dagger\mathscr{C}^{-1}\,|0\rangle = a_\mathbf{p}^\dagger\,|0\rangle = |\mathbf{p}\rangle_a \qquad (7.70)$$

so that the one-particle state is the charge conjugation of the one-antiparticle state, as it should be.

Notice that the $\mathbf{p}$ in the definition of the one-particle and one-antiparticle states is just a label, related, but not necessarily equal to the momentum carried by the state. To learn what is the three-momentum carried by such state, we must act with the momentum operator (7.45)

$$\hat{\mathbf{P}} = \frac{1}{2}\int\frac{d^3p}{2\omega_p}\hat{a}_\mathbf{p}^\dagger\,\hat{a}_\mathbf{p}\,\mathbf{S}(\mathbf{p})[\omega_p - S(\omega_p)]\frac{p_4}{\kappa} + \hat{b}_{\mathbf{p}^*}^\dagger\,\hat{b}_{\mathbf{p}^*}\,\mathbf{p}[S(\omega_p) - \omega_p]\frac{p_4}{\kappa}$$

Using the commutational relation (7.35), we find for the one-particle state (7.68)

$$\hat{\mathbf{P}}\,|\mathbf{p}\rangle_a = -\mathbf{S}(\mathbf{p})\,|\mathbf{p}\rangle_a \qquad (7.71)$$

Analogously, for the one-antiparticle state $|\mathbf{p}\rangle_b$, (7.69), using the commutational relations (7.36), we get

$$\hat{\mathbf{P}}\,|\mathbf{p}\rangle_b = \mathbf{p}\,|\mathbf{p}\rangle_b \qquad (7.72)$$

In exactly the same manner, we can use the Hamiltonian (7.44) to compute the energy of the one-particle states, obtaining

$$\hat{H}\,|\mathbf{p}\rangle_a = -S(\omega_p)\,|\mathbf{p}\rangle_a \qquad (7.73)$$

and

$$\hat{H}\,|\mathbf{p}\rangle_b = \omega_p\,|\mathbf{p}\rangle_b \qquad (7.74)$$

Therefore, one-particle and one-antiparticle states belong to the same mass-shell manifold, since

$$\omega_p^2 - \mathbf{p}^2 = m^2 = S(\omega_p)^2 - S(\mathbf{p})^2 \qquad (7.75)$$

but $\mathbf{p}$ and $S(\mathbf{p})$ are, in general different points on this manifold, with a single exception being the case $\mathbf{p} = S(\mathbf{p}) = 0$, $\omega_p = -S(\omega_p) = m$.

What we see, therefore, is that although the one-antiparticle state[4] $|\mathbf{p}\rangle_b$ carries the energy ω_p and three-momentum $\mathbf{p}$ the one-particle state $|\mathbf{p}\rangle_b$ carries the energy $-S(\omega_p)$ and three-momentum $-\mathbf{S}(\mathbf{p})$. In the limit $\kappa \to \infty$, the energies and momenta become the same, but in the case of the deformed theory they differ. However, this deformation is different from the situation of $\mathscr{CPT}$ violation considered, e.g., by

---

[4]It is, of course, a matter of pure convention, which is particle and which is antiparticle.

Greenberg in [10], because in our case, the rest masses of the particle and antiparticle are still equal, as shown above.

Finally, let us consider the mysterious fourth-momentum operator $\hat{P}_4$. It measures, essentially, the deformed charge carried by the state

$$\hat{P}_4 \,|\mathbf{p}\rangle_a = (\sqrt{\kappa^2 + m^2} - \kappa)\,|\mathbf{p}\rangle_a \tag{7.76}$$

and

$$\hat{P}_4 \,|\mathbf{p}\rangle_b = -(\sqrt{\kappa^2 + m^2} - \kappa)\,|\mathbf{p}\rangle_b \tag{7.77}$$

We see that the charge operator $\mathscr{C}$ maps the two eigenstates of the operator $\hat{P}_4$ into one another.

This completes our discussion of κ-deformed complex scalar field theory.

References

1. Arzano, M., Consoli, L.T.: Phys. Rev. D **98**(10), 106018 (2018). arXiv:1808.02241 [hep-th]
2. Arzano, M., Bevilacqua, A., Kowalski-Glikman, J., Rosati, G., Unger, J.: κ-deformed complex fields and discrete symmetries. https://doi.org/10.1103/PhysRevD.103.106015
3. Amelino-Camelia, G., Arzano, M.: Phys. Rev. D **65**, 084044 (2002). arXiv:hep-th/0105120 [hep-th]
4. Arzano, M., Kowalski-Glikman, J.: Phys. Lett. B **771**, 222–226 (2017). arXiv:1704.02225 [hep-th]
5. Arzano, M., Gubitosi, G., Magueijo, J., Amelino-Camelia, G.: Phys. Rev. D **91**(12), 125031 (2015). arXiv:1505.05021 [gr-qc]
6. Arzano, M., Marciano, A.: Phys. Rev. D **76**, 125005 (2007). https://doi.org/10.1103/PhysRevD.76.125005, arXiv:0707.1329 [hep-th]
7. Daszkiewicz, M., Lukierski, J., Woronowicz, M.: Kappa-deformed oscillators, the choice of star product and free kappa-deformed quantum fields. J. Phys. A **42**, 355201 (2009). arXiv:0807.1992 [hep-th]
8. Arzano, M., Benedetti, D.: Rainbow statistics. Int. J. Mod. Phys. A **24**, 4623–4641 (2009). arXiv:0809.0889 [hep-th]
9. Young, C.A.S., Zegers, R.: Nucl. Phys. B **809**, 439–451 (2009). https://doi.org/10.1016/j.nuclphysb.2008.09.025, arXiv:0807.2745 [hep-th]
10. Greenberg, O.W.: CPT violation implies violation of Lorentz invariance. Phys. Rev. Lett. **89**, 231602 (2002). arXiv:hep-ph/0201258 [hep-ph]
11. Sozzi, M.S.: Discrete Symmetries and CP Violation: From Experiment to Theory. Oxford University Press (2008)
12. Freidel, L., Kowalski-Glikman, J., Nowak, S.: Field theory on kappa-Minkowski space revisited: Noether charges and breaking of Lorentz symmetry. Int. J. Mod. Phys. A **23**, 2687–2718 (2008). arXiv:0706.3658 [hep-th]
13. Arzano, M., Kowalski-Glikman, J., Wislicki, W.: A bound on Planck-scale deformations of CPT from muon lifetime. Phys. Lett. B **794**, 41–44 (2019). https://doi.org/10.1016/j.physletb.2019.05.025, arXiv:1904.06754 [hep-ph]

Index

© Springer-Verlag GmbH Germany, part of Springer Nature 2021
M. Arzano and J. Kowalski-Glikman, *Deformations of Spacetime Symmetries*,
Lecture Notes in Physics 986,
https://doi.org/10.1007/978-3-662-63097-6

Printed in the United States
by Baker & Taylor Publisher Services